Introduction to Insects and their Diversity

Introduction to Insects and their Diversity

Edited by:

Laichattiwar Mukesh Anandrao

Delve Publishing

www.delvepublishing.com

Introduction to Insects and their Diversity

Laichattiwar Mukesh Anandrao

Delve Publishing
224 Shoreacres Road
Burlington, ON L7L 2H2
Canada
www.delvepublishing.com

Email: orders@arclereducation.com

ISBN: 978-1-77407-757-3 (Hardcover)

ABOUT THE EDITOR

Dr. Mukesh Laichattiwar, Ph. D (Institute of Agricultural Sciences, Banaras Hindu University, Varanasi, U.P.) is presently working as Assistant Professor at College of Agriculture, Parul University, Vadodara. Dr. Laichattiwar worked as Assistant Professor at college of Agriculture, Naigaon (Bz.), Nanded M.S. (2018-2019). He did his B.Sc (Ag.) From College of Agriculture, Ambajogai (Vasantrao Naik Marathwada Krishi Vidyapeeth, Parbhani, M.S.) and M.Sc (Ag.) from Institute of Agricultural Sciences, Banaras Hindu University, Varanasi, U.P., He was awarded UGC fellowship for his Ph. D research. He qualified ICAR-NET in 2013. Dr. Laichattiwar is a recipient of Best Poster presentation award in 2014. He Published 18 research article, 7 book chapters/proceedings, 5 popular articles and 15 abstract.

TABLE OF CONTENTS

LIST OF FIGURES

can be washed away and affect different water sources including groundwater, and they can also end up in the rain due to condensation, where they affect the crops. These are just but a few of the ways in which this happens

Figure 10.6: Certain orders have some species that only live a part of their life cycle in water, for example, megalopterans, caddisflies, dragonflies, stoneflies, mayflies, etc. while there are others that have both terrestrial and aquatic species, for example dipterans, orthopterans, neuropterans, butterflies, bugs, beetles, etc.

Figure 10.7: Above are some of the stressors causing a decline in honey bee populations. Bees are very important pollinators, and their decline is raising many concerns

Figure 10.8: Above is a representation of the only way we can end insect decline as a whole. It starts with our reasoning in a fundamental manner. We also need to make identifications of adverse synergisms and threats so that we can aim to maintain other aspects such as biomass, abundance, ecosystem services, historic communities, among others

Figure 10.9: There are many different types of pollinators; birds, rodents, monkeys, squirrels, reptiles, and even people. However, the greatest pollinators are insects; bees to be specific account for the majority of the pollinators globally. Others include butterflies, beetles, moths and wasps, all which are endangered insects

Figure 10.10: The image above shows us how the extinction of bees would affect even us as humans. 1/3rd of the food pollination is highly dependent on bees. They contribute approximately £400 million to the economy. These are just but a few reasons why bees are very important to us and to the environment

Figure 10.11: Earwigs can actually have benefits, and are considered very useful insects. They are referred to as sanitary engineers in many different parts of the world; this means that they are often helpful in cleaning up the environment as they feed on dead and live insects, as well as on decaying plant material

Figure 10.12: Above are species that are found in the Hemiptera order and, and these are referred to as true bugs. They are herbivores and have been known to cause damage to plants, which in turn causes farmers to use insecticides which can affect the insect biodiversity by killing them and other insects in the area

Figure 10.13: Insects in the Orthoptera order are considered the most dominant group in the majority of the habitats that are terrestrial. They normally feed on all plant types, therefore causing a lot of economic damage which can affect the countries economy as a whole. Parts of North America, Asia and Africa are greatly affected by swarms of locusts or grasshoppers, destroying wide land areas of crops.

Figure 10.14: Above are the four main species in this order. As stated above, they are known to be the most important pollinators globally. With all the pesticides being used, and climate change which is affecting their ideal temperatures, they are becoming endangered species, and this could affect us all

Figure 10.15: These are the common butterflies and moths we know. We see them everywhere we go, but with years, their numbers have started to drastically drop. These are also important pollinators, and their decline can be felt globally. Many of them are also highly valuable due to their beauty. Others for commerce, such as silkworms that provide silk

Figure 10.16: Dung beetles have proven to play a role that is very important and beneficial to the environment, both agricultural and natural ecosystems. Majority of these beetles use the fecal waste of many different animals as a source of food, or also for providing brood balls for larvae, which either barrow in the ground or live in chambers

Figure 10.17: Odonata is an order that includes dragonflies and damselflies. Dragonflies feed on small flying insects such as mosquitoes, and their availability is very beneficial to outdoor enthusiasts and gardeners. In addition, their presence benefits the environment by making sure that humans reduce their pesticide use for killing these stubborn insects.

Figure 10.18: There are many ways to attracts insects. One of them is by providing them with favorable habitats that consist of flowers or herbs. For example, butterflies love nectar- rich flowers and having this can attract many butterflies your way

Figure 10.19: Considering the fact that we have chosen to destroy many insects habitats, insects sanctuaries have become a modern innovation that many people are using to attract insects to the cities and towns. These sanctuaries can act as a home for insects that do not have natural habitats, especially if they are well maintained

Figure 10.20: In addition to water pollution, artificial fertilizers can degrade the soil, making it less fertile, therefore making the plants weak and unsuitable for certain insect habitats. In addition, artificial fertilizers can attract certain diseases, killing off plants or making them inhabitable.

LIST OF TABLES

PREFACE

The insect is, without doubt, the most diverse group of organisms on the planet. The research is inconclusive of the exact populations, but the estimates are between 1 and 20 million. Insects are the most abundant creatures on the earth. According to research they consist of about 75 % of all animal species already documented. Surprisingly, the species is largely overlooked despite having its existence in almost all the habitats.

Most insects are found in tropics areas and neutrophils areas countries in the Amazon regions boast many insects' species though the actual number is unknown. The majority of the human population cannot define an insect. We occasionally have heard the term bug to describe all types of invertebrates, but what is an insect? They are from the phylum arthropod, these are the invertebrates consisting of segmented bodies, jointed appendages, and continuous exoskeletons. Insects are the largest amount of this phylum and the most distinguishing differences between other species in the same group is its ability to fly.

According to research, the reproductive functions (sex) of insects are distinctive depending on the types. Some insect varieties are oviparous that is, deposition of an egg external to females' body otherwise viviparous no egg is involved in the process.

The importance of insect biodiversity in the ecosystem cannot be underestimated. They play a significant role in the ecosystem. Since they eat anything that has nutritional value, insects have successfully found themselves a niche. Some insects are scavengers who feed on the dead animals and fallen trees hence help to recycle the nutrients back to the soil. Insects act as decomposers and help to create topsoil which is a nutrient-rich layer that helps plants to grow Burrowing bugs dig tunnels that provide channels for water which then helps the plants. With their droppings, they help to fertilize the soil.

Insects also play a significant role in maintaining the food chain. For most of the amphibians, reptiles and even mammals, insects are the sole source of food for them. This makes their role in the food chain extremely important. In most parts of the world, people consume insects that are known to be a rich source of proteins, minerals, and vitamins. Some insects produce very useful substances such as wax, silk, and honey. Honey bees have been used by humans because of their honey. The Chinese also used works to produce silk which they traded with the rest of the world.

Opinions about insects are varied. For some societies, insects are feared and even harmed while in some, they are maintained. It's a fact that some insects compete for food with humans and especially in the agricultural setting. Many are also transmitters of diseases such as the female anopheles mosquito. However, it's up to us to understand that they do better than harm to us. We should all consider the overall value they place in the ecosystem.

This volume has ten chapters, and each chapter discusses a specific topic under insects and their diversity: chapter one is about introduction to insects; chapter two is about phylogeny and evolution of insects; chapter three discusses insect morphology and anatomy; chapter four discusses insect reproduction and development; chapter five presents information about insect senses and communication; chapter six discusses insect social behavior; chapter seven offers detailed information about insect locomotion; chapter eight offer information on insect ecology; chapter nine discusses insect relationship to humans; and lastly, chapter ten is all about declines in insect abundance.

These ten chapters were lengthily prepared. The author paid more emphasis on the simplicity of the content in order to make the volume comprehensible and understandable to all readers.

1

Introduction to Insects

CONTENTS

1.1 A BRIEF INTRODUCTION

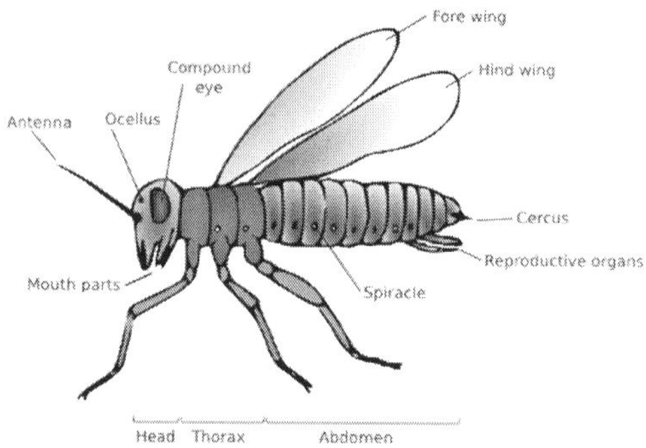

Figure 1.1: Insects have 3 body parts which are the head, thorax and abdomen.

Source: https://www.amentsoc.org/insects/what-bug-is-this/insects.html

Insects are invertebrates that form the largest set within the arthropod class. They consist of a chitinous exoskeleton, tri-part body made of the head, the thorax, and abdomen, jointed appendages, compound eyes as well as a pair of antennae. It's believed insects are the largest group of animals; covering more than 1 million different species while representing well over half of known living animals. The average amount of extant species has been estimated at roughly 6 to 10 million, possibly over 90 percent of the organism life-forms found on the planet are insects.

These creatures may be found almost in all environments, even though just a trivial number of this species live in the seas, which are domineered by another arthropod set, crustaceans. Almost all bugs hatch from eggs. Their growth is limited by the rigid exoskeleton and growth involves several molts (Capinera, 2011).

The undeveloped phases normally vary from adults in form, behavior, and habitat, plus may comprise a passive pupal phase in those groups which undergo a 4-phase metamorphosis. Bugs that undergo 3-stage metamorphosis don't have a pupal phase and adults often develop across several nymphal phases. The greater level relationship of bugs is unclear but research is still ongoing.

Adult insects normally travel by walking, flying, and sometimes swimming. Moreover, they can use a tripedal gait which allows for rapid but stable movement. In this approach, they walk having their legs in contact with the ground through alternating triangles, consisting of front and rear on one angle and middle part on the alternative side. Bugs are the only known invertebrates that have developed flight, plus all flying insects evolved from a single common ancestor.

Several insects spend some part of their livelihoods underwater, with larval evolutions that comprise of gills, plus some adult bugs are aquatic with adaptations for swimming. Other species, like water striders, can walk on water surfaces. Bugs are normally solitary, however, some, like bees, termites and ants, are social where they live in massive, well-structured colonies (Capinera, 2011).

A few insects, like earwigs, display maternal care, protecting their eggs and youngsters. Bugs can converse with one another in several ways. For example, male moths have the ability to sense female moth pheromones across vast distances. Some species also communicate using sounds, while crickets rub together their wings to attract certain mates or repel others.

1.2 KEY CHARACTERISTICS OF INSECTS

The Basic Insect Structure consists of three sets of legs including three body parts which are the head, thorax, and abdomen. Oftentimes entomologists who study bugs may also include that insects contain a pair of antennae as well as external mouthparts.

1.2.1 The Head

This region is found at the insect's body front end and comprises the mouth-parts, antennae, and eyes. Bugs have mouthparts developed to assist them in feeding on multiple things. There are insects that drink nectar and contain mouthparts adapted into a tube known as proboscis, for sucking up liquid. Some insects equally have chewing mouthparts plus consume leaves or different other plant material. Some insects bite or pinch, while others prick and suck up fluids from their hosts (Capinera, 2011).

The set of antennae might have obvious sections or resemble a feather. They are available in various forms, plus they are a clue towards classifying the bug. Antennae are applied to distinguish sounds, vibrations, or other environmental elements.

Bugs can have a pair of the compound or simple eyes. Compound eyes normally are large with numerous lenses, providing the bug a composite image of its environs. A simple eye only contains one lens. Some bugs have both types of eyes (Capinera, 2011).

1.2.2 The Thorax

It's the middle part of a bug's body, comprising of wings and legs. The 6 legs are fitted to the thorax which also has muscles that control movement. Every insect leg also has 5 parts. These legs may comprise of different shapes, including different adaptations for helping bugs to move within their unique habitat. Grasshoppers consist of legs developed for jumping, whereas honey-bees have legs having special baskets for holding pollen since the bee travels from one flower to another.

Usually, wings are available in diverse shapes and sizes, plus they are another central clue to assist you in detecting an insect. Most butterflies and moths possess wings consisting of overlapping scales, usually in bright colors. Some bug wings look transparent, with only a mesh of veins to define their shape. While at rest, bugs just like beetles or praying mantis have their wings bent flat against their frames. Other insects grasp their wings perpendicularly, such as butterflies and damselflies (Luisa, 2012).

1.2.3 The Abdomen

The abdomen constitutes the final part of the insect body which carries the insect's essential organs. Bugs have digestive organs, such as intestines and stomach, which take in nutrients from food sources and remove waste matter. The reproductive organs of insects are also present in their abdomen. Glands that secrete pheromones for tracking the insect's movement, or attracting a partner are in this area as well.

1.3 METAMORPHOSIS

Some insects like the butterfly and moth mature through a procedure known as metamorphosis. The Greek term that means conversion or changing in shape. Bugs have two main kinds of metamorphosis. Crickets, dragonflies, grasshoppers, and cockroaches undergo incomplete metamorphosis. Youngsters are called nymphs, normally looking like miniature adults but lacking wings (Luisa, 2012).

Otherwise, butterflies, moths, flies, beetles, and bees undergo complete metamorphosis. Here, the young (known as larva rather than a nymph) is rather different from adults. Similarly, it eats diverse kinds of food. Normally, there are 4 stages for the metamorphosis of moths and butterflies which are; egg, larva, pupa & adult.

Eggs are laid upon plants courtesy of adult female butterflies. These plants can then turn to become the food needed for the birthing caterpillars. Eggs may be laid during spring, summer and fall. It generally depends on the type of butterfly. The females lay several eggs at once such that at least a few of them survive.

The next phase is the larva. It's also known as a caterpillar in case the bug is a moth or butterfly. The role of a caterpillar is eating to grow. With the caterpillar growing, it ruptures its skin and disposes it around 4 to 5 times. Food taken during this period is stored and later used in adult form (Luisa, 2012).

On average, caterpillars can grow up to 100 times their scope during this phase. For instance, the monarch butterfly egg measures the scope of a pinhead plus the caterpillar which hatches from this small egg isn't much larger. However, it will develop up to 2-inches long in multiple weeks.

It then moves on to Pupa or the Transition Stage. Once a caterpillar matures and stops feeding, it turns to a pupa. Based on the species, a pupa may be suspended beneath a branch, concealed in leaves or hidden underground. Additionally, the pupa of most moths is protected within a silk cocoon.

The stage may last from just a few weeks, 1 month or longer. A number of species go through a pupal phase that lasts for nearly two years. This may seem like nothing is happening but large changes are occurring inside. Specialized cells found within the larva are currently growing fast. They shall turn to legs, eyes, wing,s and different other components of adult butterflies. Most of the first larva cells will generate energy for these maturing adult cells (Luisa, 2012).

Finally, the insect reaches the adult or reproductive stage. This level is what many people think of regarding butterflies. They appear rather different from larva or caterpillars with their tiny eyes, stumpy legs, and short antennae. Adult insects possess long legs, antennae, and even compound eyes. These can also move by using large and bright wings for flight. The adult's role is mating and laying eggs. A few species of adult insects like butterflies obtain energy through feeding on nectar though many species won't feed at all.

1.4 COMPLEX MOUTH PARTS

Insect mouthparts are structures that surround the mouth and are used for feeding, processing and manipulating food so it can easily be ingested. While functionally similar to vertebrate jaws, these lie not within the mouth, nor inside the buccal cavity.

The basic structure of mouthparts is evident in cockroaches - where they are applied for biting. Generally, there are 5 unique structures that are applied for the original collection and food processing. First, there's Labrum which is a cover that's loosely known as upper lips. They are followed by Mandibles - tough, potent cutting jaws. Them Maxillae – a pair of 'pincers' that are less powerful compared to mandibles. These are also applied to balance and manipulate food. They consist of 5 segmented palp structure that's sensory and normally used for taste (Luisa, 2012).

Additionally, there's the Labium - lower mouth cover, commonly called the lower lip. This normally represents the bonded pair of ancestral secondary maxillae. They contain a tri segmented palp that's also sensory. Finally, there's the Hypopharynx - a tongue-style structure found at the mouth floor. Moreover, the salivary glands release saliva through this.

The system remains unchanged in most insects that chew their food, larvae, and adults alike. Where unique food sources are exploited, it means the mouthparts become modified, occasionally very much, such that the food can be obtained adequately. Insect mouth components show many instances of parallel evolution, whereby there's a common end of independently achieving certain similar functions, though not across identical lines.

Many bugs consume liquid food, and this is facilitated with the creation of sucking' components from the mouth section. Honey bees, for instance, have mandibles that are rather small and appropriate for molding wax, their labium is curled downwards and inwards creating a tube appropriate for pulling up nectar. For Butterflies and Moths, these mandibles disappear completely (Luisa, 2012).

The maxillae meanwhile are elongate, structured while composed together across the surface while held together using hooks and spines for forming a sucking conduit. The proboscis meanwhile can be quite long, especially when not used where it's often coiled up to resemble a watch spring beneath the head. For diving beetle larvae, they possess a piercing and drawing mechanism.

These mandibles are curled over to create an almost fastened channel along their inward surface. Besides, the prey is often captured and pierced through the mandibles. Digestion juices are pumped across the groove. This food is turned into liquid form. The semi-digested soup would be sucked through again across the groove. For beetle larvae, these also have piercing and sucking structures. The labium structure is elongated and functions as a sheath. It covers both the maxillae and mandibles which are adapted as stylets for infiltration. The maxillae consist of two tubes that run across their length within the inward surface (Luisa, 2012).

Moreover, 'saliva' can be pumped through suction tubes. This saliva converts the food into liquid form. The liquified food consequently is pulled up through the tube. For houseflies, proboscis displays a capillary working.

This labium often is elongated and constitutes a pair of lobes at the apex. These lobes comprise of a set of well-developed tubes called (pseudotracheae). Besides, the hypopharynx stretches down the proboscis plus digestive juices move down this path onto the food. This food meanwhile is converted into a liquid with these juices. The liquid is thereafter drawn up using pseudo-tracheae through capillary action.

1.5 CLASSIFICATION

Insects are found under the taxa of > Invertebrates > Arthropods > and Hexapods >. Within these groups they are further divided into other sub-taxonomic groups which are:

1.5.1 Angel insects (Zoraptera)

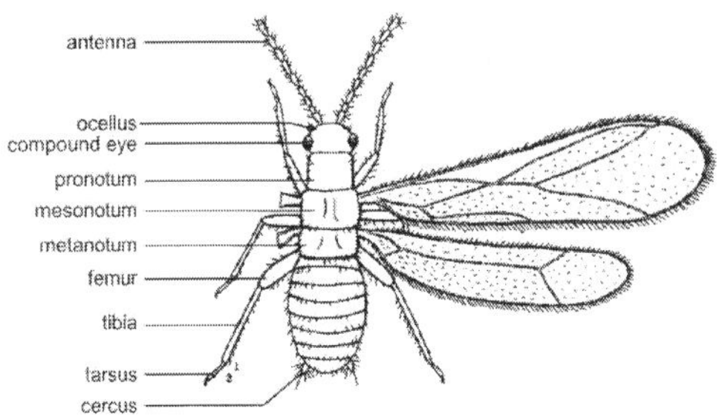

Figure 1.2: Angel insects (Zoraptera) is a small set of tiny insects.

[Source: https://link.springer.com/referenceworkentry/10.1007%2F978-1-4020-6359-6_10223]

They're infrequently encountered, and not well known. This order name is derived from the Greek terminologies zoros meaning (pure), a (without), as well as pteron (wing). On average, there are roughly 30 species defined, which all lie within the family of Zorotypidae. The angel insects are roughly 3 mm long, having a wingspan of 7 mm. They're dimorphic: a wingless-type that doesn't have eyes, ocelli, plus is only marginally pigmented, plus a winged type that contains eyes, ocelli, plus is much darker in shade (Marshall Cavendish Corporation, 2003).

Additionally, they have chewing mouthpieces. The antennae are filiform, consisting of a total of 9 segments. Their legs are unspecialized and the tarsi dual-segmented. Wings also have shortened venation, plus the wings maybe shed, just like with termites. Furthermore, the abdomen is tubular and comprises of 11 parts. Very short, single-segmented cerci happen near the abdomen tip, while metamorphosis isn't pronounced.

When this group name was first proposed, entomologists thought that zorapterans were entirely wingless. Afterward, it was found that a number of zorapterans are winged, plus that winged adult insects cast-off their wings. Besides, adults are normally small-bodied, having a hypognathous head plus chewing mouthparts. Their antenna is moniliform as highlighted, but the nymph stage has 8 antennal segments.

The zorapterans wings are membranous with lessened venation and are normally shed from the bottom. Wingless forms don't have ocelli or compound eyes; while winged insects have both. Legs exhibit tarsi with a pair of sections. The abdomen consists of 11 sections, the ovipositor is stunted or absent, while cerci are existent but not segmented (Marshall Cavendish Corporation, 2003).

Growth is gregarious, typically under planks, in heaps of old sawdust, rotting logs, under barks, or in relation to termites. These insects are mostly fungivorous or necrophilous, sometimes both. Only a pair of species are available in North America; occurring from Iowa to Pennsylvania and even Texas and further eastwards.

Their makeup is simple, consisting of biting, generalized mouth-parts, plus 5-sectioned maxillary palps and tri-segmented labial palps. Occasionally both genders are apterous, while in some forms the rear wings are tinier than the front wings; besides the wings are normally shed like in ants or termites. The wing venation equally is highly specified and compacted (Marshall Cavendish Corporation, 2003).

Originally, the order comprised of just one family called (Zorotypidae) and the genus (Zorotypus), ,however, it has since grown to include a total of 7 genera delimited primarily on wing venation. But the subdivision of Zorotypus isn't accepted by certain scientists. The phylogenetic status of Zoraptera depending on morphology has furthermore been controversial, varying through the association of the hemipteroid categories, Isoptera, orthopteroid, or the blattoid.

Analyzing major wing structures or musculature shows that Zoraptera falls in the blattoid family. Even though their wing-shape and venation look like the ones for narrow-winged Isoptera, still cephalic and abdominal features show an early deviation from the blattoid pack, before the Dermaptera divergence, and way before the beginning of the Dictyoptera family line.

Normally, *Zorotypus gurneyi* reside in colonies comprising of even several 100s of individuals. More generally the colonies consist of a size of roughly 30 individuals, out of which roughly 30% are nymphs, while the rest are adults (Marshall Cavendish Corporation, 2003).

The Zoraptera spend the majority of their moments grooming each other. The grooming procedure is believed to be an inventive way of eliminating fungal pathogens. Whenever a pair of *Z. hubbardi* colonies are pooled together experimentally, there's no big difference in terms of behavior towards other members of the same or new colony.

Thus, colonies found in wild areas may combine easily. Winged varieties are rare. Males in an average colony develop a linear dominance order, where age or colony duration membership is a prime element determining dominance. The males showing later in colonies rank at the very bottom of the structural ladder, notwithstanding their body size. Through continually attacking different males, the domineering male controls a group of females. All members of this female harem remain clumped together.

There's a strong connection between the rank and propagative success of males. While *Z. barberi* don't have such a control structure, they nevertheless display complicated courtship behavior such as nuptial feeding. Most males comprise of a cephalic part, which opens at the center of their head.

In courtship, they discharge a liquid from this gland, plus provide it to females. The reception of this droplet coming from the female serves as a behavioral releaser and instantly delivers copulation. For *Z. impolitus*, copulation doesn't happen, but fertilization is done rather by the transference of a spermatophore donated from the male insect to female (Marshall Cavendish Corporation, 2003).

The 0.1-millimeter spermatophore comprises of one massive sperm cell, unraveling to roughly the same length compared to the female herself, which is 3 millimeters or (0.12 in). It's believed that this massive sperm cell inhibits fertilization by different males, through physically jamming the female's genital parts.

1.5.2 Barklice and Booklice (Psocoptera)

Figure 1.3: Barklice are stout and stocky insects with large antennas.

[Source: http://nathistoc.bio.uci.edu/Psocoptera/index.htm]

While they're indeed called lice, Psocoptera is free-living bugs, not parasites. This scientific name is derived from the Greek word "psocus" which means (to grind), pteron (wing) referring to the psocopteran mandibles, which are designed to grind food, more like a mortar &pestle. The insects may be convenient, if rather arbitrarily, mentioned in two groups, which are booklice and barklice, depending on whether they likely reside outdoors or within human habitations. Many of them stay outdoors.

The barklice are normally found in humid places, like leaf litter, under stones, on foliage, or beneath tree bark. These have elongated antennae, wide heads and 'protruding' eyes. They live off algae, fungi, and lichens, including different plant products, like pollen. These insects can grow to 10(mm) in length, but they're normally less than 6mm, plus adults are frequently winged. The wings often are supported roof-like above their bodies (Resh & Cardé, 2009).

The winged barklice resembles an aphid, only for their elongated antennae, broadened heads and piercing jaws reveal that they aren't similar. There are certain species that are gregarious, residing in small colonies under a gossamer blanket rolled with silk coming from labial glands found in their mouth. Remarkably, both adult barklice and their larvae can produce silk from these particular glands. At times, the colonies look like they're moving in a coordinated manner, more like ants (Resh & Cardé, 2009).

For booklice, they are wingless and way smaller compared to barklice (lesser than 2 millimeters). They are typically found in human habitats and warehouses. Most of them belong to the family of Trogiidae. A number of booklice classes eat on stored grains, book bindings, wall-paper paste or other starchy products, plus on the little traces of mold present in old books.

These can sometimes result in damage to kept museum gatherings of bugs and plants. Some few, like *Liposcelis decolor*, may even be a pest found in grain stores. Most indoor residing Psocoptera have bad eyesight, plus some even seem to converse using sound. These species tap using the rear of their abdomens to produce a faint knocking noise, thus they have been named 'ticking spiders'.

The *Liposcelis divinatorius* is a good example of the 'ticking spider' – even though it's an insect and not spider. A number of Psocoptera also reside in bird nests. Particularly, those which this feed on remains of skin cells or feathers, not on the birds in particular. Psocoptera undergo incomplete metamorphosis (that is, they lay eggs, which grow into nymphs and adults) plus the order is subdivided into 3 other suborders which are Psocomorpha, Trogiomorpha, and Troctomorpha

They are distinguished with the amount of segments found in their antennae, tarsi, plus labial palps. The Psocoptera are considered the most primeval amongst the hemipteroids (real bugs, thrips, and lice) since their mouthparts reveal the least adaptation from those of earlier known fossils (Resh & Cardé, 2009).

For Phthiraptera, or (parasitic lice), it might have evolved straight from commensal barklice, plus a close association between barklice and other parasitic lice is further supported by comparisons in the arrangement of their mouthparts. Generally, there are more than 1,000 types of Psocoptera insects in Europe, with just about 100 in the British Isles alone. The Global Catalogue published by Geneva Natural Histories Museum lists around 41 families, 371 genres and 4,408 species according to worldwide estimates, going up to the year 2000.

Outdoor species can be found using sweep nets or beating trays, plus by looking beneath the bark or some older birds' nests. The Indoor species may be found through searching in musty-smelling, abandoned corners - including, possibly, even your most prized bug collection (Capinera, 2011).

1.5.3 Bees, Ants, and their relatives (Hymenoptera)

The most common Hymenoptera are bees, ants and wasps; bugs are feared by most individuals due to their stings. Nevertheless, they constitute just a small number of the order. Hymenoptera also comprises of Ichneumons including other parasites like "wasps" and Sawflies. Regrettably, there's no English term to embrace this entire order like there is for bugs, beetles and real flies.

Figure 1.4: Bees, ants, and their relatives (Hymenoptera).

[Source: https://www.pinterest.com/pin/464785624028330982/?autologin=true]

Hymenoptera naturally has 2 pairs of wings, massive fore pair plus a smaller back pair. The wings are bound together by a set of hooks (known as the frenulum) plus may resemble a single pair for the naked eye. These creatures also tend to possess prominent antennae, normally with 9 or more sections and biting mouth-sections. In Britain alone, there are roughly 6,700 Hymenoptera species, found in all land habitats. Within the global scale, there are roughly 100,000 species (Resh & Cardé, 2009).

For pollination, many winged Hymenoptera normally visit flowers. Bees perform this to gather pollen and nectar sap for themselves or giving their maturing young, whereas wasps are typically on the watch out for other bugs to feed on.

From an evolutionary perspective, flowering plants have largely taken advantage of this factor by providing nectar to bugs most likely for carrying their pollen to a different flower. With most important plants reliant on insect pollinators, it means Hymenoptera has a big role to play towards feeding the global human population. Moreover, bumblebees are normally introduced into industrial greenhouses whereby pollinators are in limited supply, and the additional pollination contributes to the globe's economy is billions of dollar-years.

Maturation: Hymenoptera passes through a full metamorphosis -meaning, they have egg, larval, pupa and adult phases. The larvae are normally soft and white having no legs, resembling Diptera larvae; besides the larvae of a few Sawflies have tiny fore and hind limbs like caterpillars. The genders are different, though males are unique in various Parasitica, and the genders are "restructured" a bit in social aculeates. A number of Hymenoptera are equally protandrous, meaning the adult male emerges first before the female while waiting for the appearance of females to reproduce with or mate before or during emergence. Oppositely, females of a few Parasitica groups shall wait for their male offspring to appear before quickly mating with them (Resh & Cardé, 2009).

1.5.4 Beetles (Coleoptera)

It's the biggest order in the Insecta class with more than 250,000 defined species. Evidently, with this high number of species, there's a broad diversity in this group with regards to size, physical features, biology, and behavior. Nevertheless, the order is normally marked by a set of elytra (toughened front wings), including a set of membranous rear wings.

These insects have chewing mouthparts which can be slightly adapted for different functions, and pass through full metamorphosis.

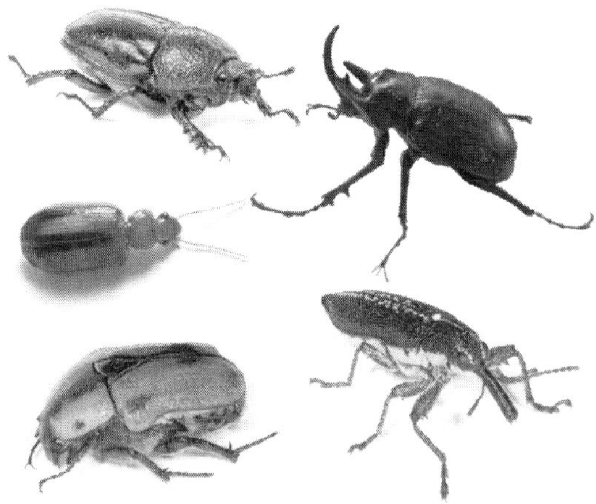

Figure 1.5: Beetles are found in almost every kind of habitat, ranging from deserts to marine systems, normally feeding on different plants and animal matter.

[Source: https://en.wikipedia.org/wiki/Beetle]

Types of Beetles (Coleoptera)

- Leaf Beetles (Chrysomelidae). They are phytophagous by nature, meaning the insects are oval, plus normally brightly shaded. Their antennae are noticeable, though characteristically less than half long as their body. The tarsal formulation (amount of tarsomeres found on fore, central, and back leg) is 4-4-4, but some are 5-5-5. They're the same as long-horned beetles, apart from their antennae which are normally smaller while their eyes also aren't notched.
- Lady Beetles (Coccinellidae). These beetles are normally recognized by their unique coloration and oval-shaped body shape. Their head is normally partially or wholly concealed with the pronotum. Additionally, their antennae are normally short with around 3- to six-segmented clubs. Additionally, their tarsal formula is outwardly 3-3-3, however truly 4-4-4. Many adults and

immature lady beetles generally are predators, where they have been used in biological control activities. Their adults normally overwinter in huge aggregations.

- Scarab Beetle (Scarabaeidae). These beetles are moderately stout-bodied and vary immensely in size. There are around 1,400 scarab species available in North America, most of which cause significant damage to different plants. Their tarsi are altogether five-sectioned (5-5-5). The antennae also have 8-11 sections and are lamellate, allowing them to hold onto the sections of the club narrowly together or expanded. Some common scarabs are the May-June beetles, Japanese beetles, and masked chafers.
- Tiger Beetles (Cicindelidae). The adult tiger beetles normally are metallic with unique color patterns. Their unique shape is normally used in classifying them. Their head is usually as broad as or bigger than the pronotum. Besides, they have thread-like antennae and long slender legs. The insects are quite fast and regularly difficult to gather. Their agility and robust mandibles make them nicely-adapted for their predation lifestyle. The threatened Salt Creek Tiger-Beetle is a part of this family. In addition, authorities place tiger beetles within the same class as ground beetles (Carabidae).
- The Ground Beetles (Carabidae). These carabids are quite similar to tiger beetles and often considered to be grouped in the same family. They can be identified by having heads that are narrower compared to the pronotum. Besides, they are normally black and shiny, however, some species are radiantly colored. Just like tiger beetles, these have lengthy slender legs. Almost all species are predatory as both larvae and adults.
- Longhorned Beetles (Cerambycidae). These beetles are recognized by having antennae which are ordinarily at least as half as lengthy as the body plus often longer. Their body is normally cylindrical and lengthened. The eyes are frequently jagged around the antennae. Every long-horned beetle is phytophagous, they are primarily wood-boring when in the larvae stage and flower-feeding when adults. Their tarsi are alike to chrysomelids, numbering 4-4-4.

1.5.5 Bristletails (Archaeognatha)

Figure 1.6: A camouflaged Bristletail resting on a leaf.

[Source: http://tolweb.org/Archaeognatha/8207]

Estimates indicate that there are roughly 350 unique species of bristletails in existence today. The members of this group don't go through metamorphosis (young bristletails resemble tiny versions of their parents). Bristletails have a tubular body that forms to a tapered bristle-like tail.

Characteristics: The Archaeognatha are part of the Thysanura order of insects which on average measure 6 to 25 mm in length. They have large and contiguous eyes, with a pair of lateral and a median ocellus (tiny single eyes with one beadlike lens). Their flagellate antennae extend approximately 1/2 to 3 times the body length. The mouth bits are ectognathous (freely visible) while the mandibles are attached with the head through a monocondylic joint (that is, a point of attachment).

Moreover, some scientists believe that the feature differentiates Archaeognatha from other ectognathous Insect groups. The 7-segmented maxillary palps aren't longer than their legs for comparison. Besides, the thoracic tergites II and III are in lateral angle strongly arched, while the 2 or 3 tarsal sections of the legs are firmly united (Shaw, 2014).

Some taxa have extra scopulae (solid brushes of specified hairs) on their distal end of the 3rd tarsal section. There are typically pairs or double sets of eversible vesicles present on the abdomen coxites. On every one of the abdomen coxites II up to IX, styli are existent.

Their two massive compound eyes merge at the uppermost part of the head, plus there are a total of three ocelli. The insects' mouthparts are especially retractable, with basic chewing mandibles and lengthy maxillary palps.

Additionally, Archaeognatha vary from Zygentoma in different ways, like their comparatively small head, and bodies flattened laterally (from one side to another) rather than compressed dorsoventrally, plus in their ability to apply their tails to jump up to 30cm into the air when disturbed. They likewise are unique among bugs in possessing tiny, articulated "styli" on their hind (and occasionally middle) coxae plus on the sternites 2-to-9, which some experts regard to be vestigial attachments.

They have double reversible membranous vesicles by which they intake moisture. Their body is also covered with easily detached scales, which make the animals hard to grip and may safeguard the exoskeleton from scraping. The thin exoskeleton provides little protection if any against dehydration. The insects accordingly should remain in humid air, like in cool, damp conditions under rocks or bark (Shaw, 2014).

As for habitat, they can be found mostly in moist soil, though some have evolved to chaparral, including sandy deserts. They survive mostly on algae, however, also mosses, lichens, and decaying organic debris make up their diet.

1.5.6 Caddisflies (Trichoptera)

Figure 1.7: Caddisfly with long antennae and wings.

[Source: https://www.aquatax.ca/trichoptera.html]

On average, there are roughly 7,000 species of caddisflies in existence today. The members of this set have aquatic larvae that develop a protective casing where they live. The casing is developed of silk developed by the larva, plus also integrates other materials like organic debris, twigs and leaves. Their adults are nocturnal, plus short-lived (Shaw, 2014).

Nevertheless, it's the biggest order of insects that's fully aquatic. They can exist in all kinds of lotic (streaming water), and lentic (still water) habitats, such as temporary water bodies. The capacity of caddisflies to stay in these diverse habitats and the immense diversity of species may be credited to their usage of silk threads to produce flexible or attached cases.

These cases normally are made from rock, wood, or detritus. They provide safety and camouflage within the ecosystem where they live. Caddisfly larvae look like caterpillars with fused legs and may be differentiated from other aquatic bugs by their toughened head capsule, small antennae, no wing pads, and softened abdomen. Adults have a pair of long antennae and wings enclosed with rather short hairs held bent, or roof-like just above their abdomen (Shaw, 2014).

These insects usually have a dull appearance (some are uniquely patterned) with lengthy, slender antennae. They have chewing type mouthparts containing reduced mandibles, however, feed mostly on liquid diets when grownups and are predatory or detritivorous as larvae.

Grownups are nocturnal and usually reside alongside lakes or streams, however, they can also be found in other places apart from water bodies. Larvae can be found in various aquatic habitats, mostly cool quick-flowing streams. Caddisflies are a crucial part of aquatic food chains plus also is applied as pollution biosignals. Additionally, caddisfly larvae can easily be differentiated from other water insects by the existence of adapted rear prolegs and a flexible case or silky retreat that shields their soft body (Cranston & Gullan, 2010).

Features: Adult caddisflies are usually 3 to 15 millimeters in length. Their frontal wings typically vary from 4 to 20 millimeters in length, having wingspans of roughly 8 to 40 millimeters. At rest, the wings are bent roof like and protect the upper part of the body. One particular family, commonly referred to as microcaddis, measure just 1.5 millimeters in length and have anterior wings spanning 2 to 5 millimeters.

Caddisfly wings either consist of hairs all over their bodies or have hairs covering their veins. The back wings are usually broader compared to their anterior wings. These insects are widely dispersed in freshwater habitats all over the world. Additionally, the larva of a particular marine species, known as *Philanisus plebeius*, can be found in intertidal zones across New Zealand and the southern Australian shores.

Besides, the larva of a land -based European species, called *Enoicyla pusilla* resides in damp leaf litter, whereas other species are found in brackish water. Roughly 7,000 species of all existing caddisflies are known (Cranston & Gullan, 2010). They lay large quantities of eggs numbering even up to 800, these are laid inside a jellylike substance that swells upon coming into contact with water. The female insect may wash away a partly extruded egg mass through dipping its abdomen into the water amid flight or place the eggs on stones inside the water or aquatic plants with surfaces just over the water. Young larvae hatch in just a few days. Based on the specific species, larvae can be carnivorous, herbivorous, or even omnivorous. For some species, their larvae form debris networks for safety, while others build a funnel-like system between stones in flowing water to capture food.

Some insects guard their bodies using special cases, while others weave protective shields or live freely. They weave silk from glands found on the lower mouth (labium), while many herbivorous species create tubular protective cases that are open at either end and expand as the larvae mature (Cranston & Gullan, 2010).

1.5.7 Cockroaches (Blattodea)

These are insects under the order of Blattodea. They are majorly known to live around the human environment.

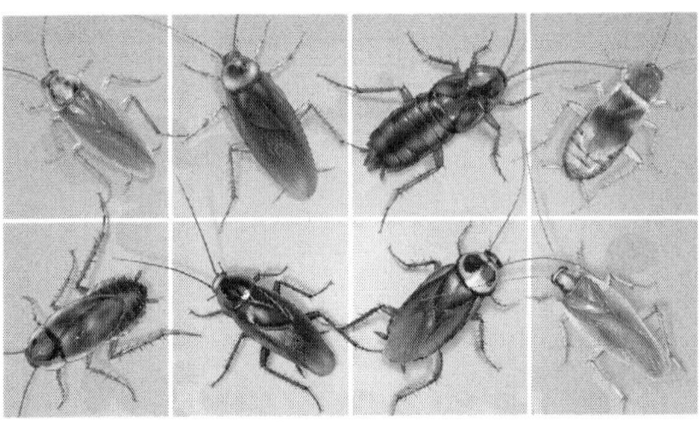

Figure 1.8: There are various kinds of cockroaches more so for the household insects which include the German cockroach, American cockroaches, and Australian cockroaches and there is the oriental cockroach.

[Source: https://cockroachfacts.com/types-of-roaches-with-pictures/]

There are various families of cockroaches. They are said to be insects that lack certain adaptations which include sucking mouthparts of aphids among other insects. Cockroaches are insects known to survive in a variety of environments that range from arctic to tropical environments. There are some species of insects that have an elaborate social structure that include a common shelter, kin recognition, social dependence and information transfer. Cockroaches have a broad variety of habitats in the areas of the world (Berenbaum, 2010).

When discussing the taxonomy and evolution we get to understand that cockroaches belong to the order of Blattodea. The name Blattodea is a scientific name that was derived from a Latin blatta which means an insect that shuns light. This name was also to mantids. In some instances, the term Blattoptera has been used to refer to extinct cockroaches. Most of the species of cockroaches are about the size of a thumbnail while some of them are bigger than that. Insects have a small head and broad, flattened body, and their color may vary from reddish-brown to dark brown. Their mouthparts are fitted inside the head. The mouthparts include chewing mandibles, salivary glands, and also taste and touch receptors.

The body of cockroaches is divided into a thorax of three segments and a ten-segmented abdomen. The external surface of the cockroaches is made of an external surface contains a tough exoskeleton having calcium carbonate and it protects the inner organs. It is also an area that provides attachment to muscles. The tough exoskeleton is coated with wax thereby repelling water. There are wings attached to the second and third thoracic segments. There is the first pair of wings also known as the tegmina are tough and protective and it lays as a shield on the membranous hind wings which are utilized in flight. The four wings have branching longitudinal veins and also multiple cross-veins.

It also has three pairs of legs which are sturdy and have large coxae and five claws on in the individual legs. The legs are attached to the three thoracic segments; the front legs are usually the shortest while hind legs are the longest giving cockroaches the main propulsive power when the insects are in fast motion. It was initially considered that the legs are sensory. The wire meshes present on the legs were discovered to aid locomotion in difficult terrain. Their abdomens have ten segments each of which has a pair of spiracles used in respiration. The tenth segment contains a pair cercus, a pair of anal styles, the anus and the external genitalia. For the male species, they are aedeagus through which they secrete sperm in the process of copulation while, the females have the spermathecal which they utilize in the storing of sperms. They also have the ovipositor which allows the ootheca is laid.

1.5.8 Crickets and Grasshoppers (Orthoptera)

The grasshoppers and crickets belong to the order of Orthoptera. This order is divided into two include Caelifera and grasshoppers are under this sub-division along with other insects. The other subdivision is Ensifera where crickets and its relatives belong. The insects in these orders are known to undergo incomplete metamorphosis. They also produce sound a process known as stridulation which is made possible by them rubbing their wings or legs against each other. Their wings and legs contain rows of corrugated bumps. Their ears are located in the front of the tibia in crickets such as mole crickets. For the case of grasshoppers and locusts, the ears are located in the first abdominal segment.

The organisms in this order make use of vibrations in the location of other individuals. These groups of insects can fold their wings. Some of the characteristics of these insects include them having a cylindrical body. They also have elongated hind legs and musculature that are adapted for jumping. They also have mandibulate mouthparts used in biting and chewing. They have large compound eyes and may not have ocelli depending on the species. They also have antennae that have multiple joints as well as a filiform type. The antennae have a variable length.

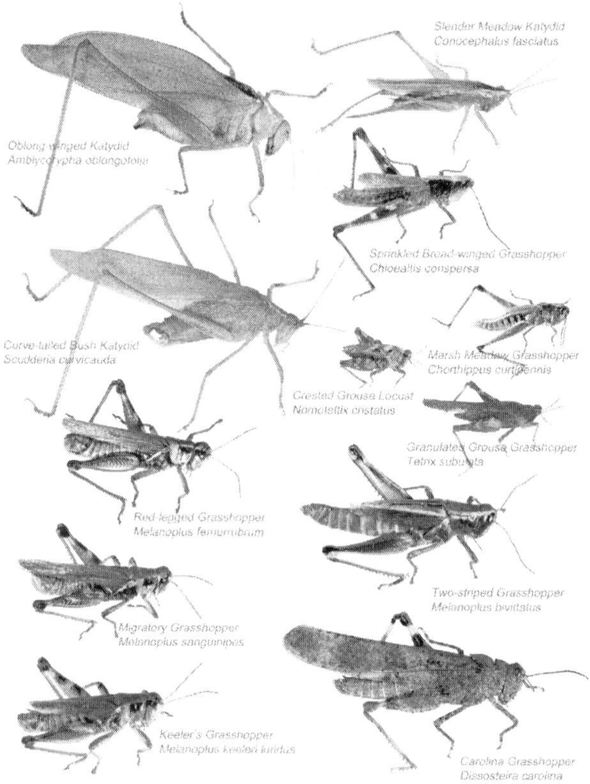

Figure 1.9: Different types of cricket and grasshoppers.

[Source: https://ofnc.ca/ofnc-event/grasshopper-outing]

For these insects, the first and third segments of the thorax are larger. The second segment is smaller. They also have two pairs of wings that are overlapping when at rest. The tegmina are narrower than the hindwings and are hardened at the base. The hindwings are membranous and have straight veins and numerous cross-veins. While at rest, the hindwings are held folded fan-like and under the forewings. For the last two and three segments of the abdomen are usually reduced and have a single segment. In the case of life these insects have an incomplete metamorphosis. Sound is an important element in the courtship process. The songs are distinct crickets have their sounds and grasshoppers have their sounds. Most of the grasshoppers lay their eggs on vegetation or on the ground. The eggs hatch and moult into nymphs that resemble the adults but do not have wings.

While in this stage they are referred to as hoppers. They may possess radical different coloration from the adults. As the moulting possesses continues the nymphs develop wings until the final moulting process where they become fully mature and have fully developed wings. The number of moulting varies depending on the species. Other factors such as growth also vary but other factors come into play such as the amount of food that was available at the moment and also the weather conditions. In some cases, these insects may be kept as pets as they tend to feed on some crops of farmers. They can also be utilized as food or as creators of biofuel.

1.5.9 Damselflies and Dragonflies (Odonata)

These insects belong to the order of Odonata also known as toothed ones. In this order, there are three groups which include the anisoptera which involves the dragonflies, the zygoptera including the damselflies and the anisozygoptera.

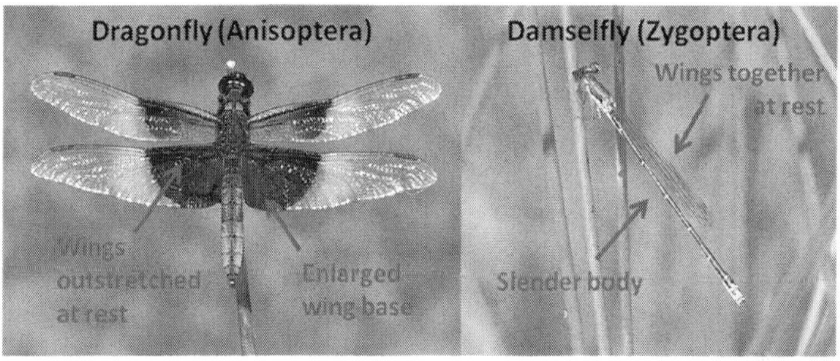

Figure 1.10: Damselflies and dragonflies (Odonata).

[Source: https://ovlc.org/dragonflies-and-damselflies/]

Dragonflies and damselflies are under the subgroups of insects and are therefore are in the group of uniramian arthropods. There are many of the characteristics that make them be unique groups of insects which include minute antennae. They also have extremely large eyes which fill most of the head, two pairs of transparent membranous wings have small veins. They also have long slender abdomen as well as an aquatic larval stage which is from the nymph with posterior tracheal gills and prehensile labium which have an extendible jaw underneath the head.

The distinguishing characteristic between dragonflies and damselflies is the size of the eyes and the shape of the abdomen. In the case where the insect in question has eyes that are relatively large in proportion to the head and the abdomen is long and thin then the insect is under the order of Odonata. Dragonflies and damselflies share many common features but there are distinct features. There is a difference in the morphology of the eggs that are used in distinguishing between damselflies and dragonflies. In the case of the dragonfly's eggs, their eggs are round and about 0.5 mm long while in the case of the damselflies their eggs are cylindrical and longer with the eggs having a length of 1mm long. Also in the case of the nymphs' stages, there is a difference. In the larval stage, the damselfly abdomen is longer and narrower.

They also have three fin-like gills projecting from the end. In the case of the nymphs of the dragonflies, they are shorter and bulkier while the gills are located on the inside of the abdomen. These nymphs expand and contract in the abdomen to facilitate the movement of water out rapidly to allow for a short burst of underwater jet propulsion. For the dragonfly, most of its life is spent in the larval stage with it undergoing molting from six to fifteen times. This is dependent on the altitude and latitude. The development stage occurs the nymph crawls out of the water and molt for the last time. By so doing it emerges from its old skin as an adult functional wings. They lack an intermediate pupal stage before they become adults. This results in them being referred to hemimetabolous as they undergo incomplete metamorphosis.

In the case of these insects, they have large heads with compound eyes which are large. Individual compound eyes have individual units of 28000 and along with other eyes cover most of its head. This implies that about 80% of their brains are utilized in analyzing of the visual information. Their antennae are tiny while their mouths have been adapted for biting hence they become efficient hunters. They have a prehensile labium that has been extended forward and underneath the head.

1.5.10 Earwigs (Dermaptera)

These are insects under the order Dermaptera. There are twelve families of insects under this order. In the case of the Earwigs, they have unique cerci and a pair of forceps that are like pincers on their abdomen and membranous wing that are folded underneath forewings that are short hence the name skin wings.

Figure 1.11: Earwigs (Dermaptera).

[Source: https://news.mongabay.com/2014/11/gone-for-good-worlds-largest-earwig-declared-extinct/]

In the case of the Earwigs, they are found in the continent of Antarctica. They are said to be nocturnal and hide in small moist crevices during the day. They are usually active during the night. They feed on plants and insects. When it comes to the foliage flowers the earwigs are usually blamed for the damages that are incurred.

For these insects, they undergo five molts before they become adults. These species of insects are unique in that they display maternal care which is not demonstrated by other insects. The female lays eggs and takes care of them even when they hatch as nymphs will continue to watch over the nymphs. The care is done until the second molting process. While the molting process occurs the sexual dimorphism begins to be more visible. Most of the earwigs are flattened allowing them to fit in certain areas. Their bodies are elongated with the largest species being the Australian giant earwig.

These kinds of insects are usually characterized by the cerci which are referred to as a pair of forceps-like pincers on their abdomen. In the case of the male earwigs generally have more curved pincers compared to the females. The pincers are used in the capturing of prey and also when they need to defend themselves and fold their wings under short tegmina. They have antennae which are thread-like having a minimum of ten segments. The forewings are usually short oblong leathery plates that cover the hindwings like the elytra of a beetle. Their wings are not used in flight. Most of the species have short and leather like-forewings that have thin hindwings.

The hindwing is a thin membrane that expands like a fan that radiates from one point which is folded under their forewing. Though most of the species of insects have wings and have the ability to fly, they rarely utilize their wings in flight. Their wings usually have a unique venation and in the pattern of folding that requires the use of the cerci. Their life cycles involve the females laying eggs and they develop through an incomplete metamorphosis which may go through 4 to 6 molts. There are developmental stages present between molts known as instars and live a year from hatching. The males and females live in chambers in debris, crevices or in the soil. After mating the sperms may remain in the female for months before the eggs are fertilized. When the legs are first laid, they are usually white or cream-colored and oval-shaped. Before they hatch they become kidney-shaped and are brown.

1.5.11 Flies (Diptera)

They are insects under the order of the Diptera. These kinds of insects make use of a single pair of wings to fly. The hindwings have evolved and have become advanced mechano-sensory organs referred to as halters that act as high- speed sensors of rotational movement they allow the dipterans to perform advanced aerobatics. Files have a mobile head and have a pair of large compound eyes. Their mouths have been modified in piercing and sucking and in some instances are used in lapping and sucking. Their wings are arranged in a manner that allows easy movement while they are in flight. There are claws and pads on their feet that allow them to cling to smooth surfaces. In the case of insects, they undergo complete metamorphosis. The eggs are laid on the larval food-source and the larvae.

They lack true limbs that develop in a protected environment and some cases in their food source. While in the pupal stage it is usually a tough capsule from which the adult emerges. In this case, flies tend to have a short life as adults. There are other insect orders but Diptera is the major order. They are really important when it comes to pollinators after bees. Fruit flies are utilized in a model organism in matters of research. Their buzzing sounds are termed to be an annoyance to most people around the world as they tend to settle on the skin or eyes. There are larger species of flies which include tsetse flies.

Flies are found in the terrestrial habitats in the world apart from Antarctica. These insects are adapted for an aerial movement and tend to have short and streamlined bodies. There is the first tagma of the fly. There is the head which carries the antennae and mouthparts that include the labrum, labium, mandible, and maxilla. There is the second tagma which is present in the thorax which carries the wings. It is also the area that carries the muscles that are utilized in flight. This segment is greatly enlarged. The first and third segments are reduced to collar-like structures. The third segment has the halters which aid the insects in the process of flight.

In the third tagma is the abdomen that is made up of 11 segments which have been fused. The three hindmost legs have been modified for reproduction. In the compound eyes have three ocelli on the top. The compound eyes may be located close together and in some instances, they may be widely separated. They also have antennae that have been well developed but vary and are thread-like, feathery or comb-like in a variety of families. Their mouthparts have been modified to allow piercing and sucking. To facilitate visual control, the flies have an optic flow. They have one pair of fore wings on the mesothorax and a pair of halters and have reduced hindwings on the metathorax. Other adaptations have been made to them to facilitate flight.

1.5.12 Mantis (Mantodea)

This houses insects such as mantis. They are found in temperate and tropical environments. They have triangular heads and have bulging eyes that have been supported by flexible necks. They have elongated bodies that may or may not have wings. They have forelegs that are greatly enlarged and have been modified to catching and gripping prey. They have an upright posture. Their closes relatives are the termites and cockroaches. They are known to be ambush predators. There few of them that ground-dwelling predators and are said to be actively pursuing their prey. The adults lay their eggs in the autumn then they die. Their eggs are protected by their hard capsules and hatch in the spring. The females are said to practice sexual cannibalism and eat their mates during the process of copulation.

Figure 1.12: Mantids (Mantodea).

[Source: https://en.wikipedia.org/wiki/Mantis]

When we look at the anatomy of the insects, they have large triangular heads and have a beak-like snout and mandibles. They have two bulbous compound eyes. They also have three small simple eyes and also a pair of antennae. There is an articulation on the neck that is flexible. Some of the preying mantis could rotate their heads an angle of 180. The thorax of the mantis is made up of a prothorax, a mesothorax, and a metathorax. The prothorax carries the head and the forelegs which is much longer than the other two thoracic segments. It is also flexibly articulated facilitating a wide range of movements of the head and the forelimbs while the remaining part of the body is more or less immobile.

They have two spiked, grasping forelegs which facilitate prey being held securely. In this case, the coxa and the trochanter combine resulting in the formation of a segment making it to be as long as the femur which is a spiky part of the grasping apparatus. At the base of the femur is a set of discoidal spines that in most cases occur in four it may range to five depending on the type of species. The spines are then preceded by a number of tooth-like tubercles along the tibia and the apical claw near its tip. The foreleg is then able to grasp better. The foreleg usually ends in a delicate tarsus and is utilized as walking appendages which are made of four or five segments and end in two-toed claw with no arolium. These kinds of insects have a stereotype of vision which allows them to locate the prey. They make use of their compound eyes. They are generalist predators of arthropods. They are said to be ambush predators that feed on prey that is greatly available. In some cases, the large mantises feed on the smaller ones.

They are also preyed on by the vertebrates and to avoid this their bodies have been adapted to allow them to escape from these predators. For them, they mate in temperate climates more so in autumn.

1.5.13 Mayflies (Ephemeroptera)

These are aquatic insects belonging to the Ephemeroptera. The mayflies are also known as Canadian soldiers. They exhibit aspects of ancestral traits. In the case of the immature mayflies, they are usually aquatic and are said to be nymphs. The adults tend to live short lives and the nymphs usually spend most of their time underwater. The insects have an elongated, cylindrical, and flattened body that undergoes a number of instars, molting, and increasing in size each time. When they are ready to emerge from water, the nymphs usually vary in length depending on the species.

Figure 1.13: Mayflies (Ephemeroptera).

[Source: http://www.cirrusimage.com/ephemeroptera_mayflies.htm]

Their heads have a tough outer covering of sclerotin. They have various hard ridges and projections that point downwards. They have a mouth at the front. They have two large compound eyes and have three simple eyes. They also have antennae of variable length and are set between or in the front of the eyes. Their mouthparts have been modified to allow chewing and they are made up of a flap like-labrum, a strong mandible, a pair of maxillae, membranous hypopharynx, and labium. When looking at their thorax, they are made of three segments having the mesothorax and the metathorax that have been fused. The segment has a pair of legs that live in a single claw. Their legs are robust. They often clad in bristles, spines or hairs. Their wing pads develop on the mesothorax while in other species there is the metathorax.

The abdomen is made of ten segments which may be obscured by a large pair of operculate gills, thoracic shield or the developed wing pads. In the case of the subimago, this is the stage where the nymph resembles the adult. They are usually sexually immature and are duller in color. In the subimago, there are the partial cloud wings fringed with minute hairs known as microtrichia whereby its eyes, legs, and genitalia are not fully developed. The female species may lack the subimago state the adult stage appears like a subimago with the microtrichia on the wing membrane. While in this state they are poor fliers and lack color patterns that may be used to attack mates. This is after a period that lasts one or two days while in some species it lasts a few days.

In the case of the adult mayflies, they are relatively primitive in structure and possess traits that were possibly present in the first flying insects. They also have long tails and wings that do not fold over a flat abdomen. They are usually delicate- looking insects that have two pairs of membranous triangular wings that are extensively covered with veins. When at rest, the wings are held upright. The hind wings are much smaller compared to the forewings and the vestigial may be absent. The second segment in the thorax bears the forewings. They are enlarged to hold the main flight muscles. The adults have short flexible antennae, large compound eyes, non-functional mouthparts, and three ocelli. In most cases, the male eyes are larger while the front legs are longer.

1.5.14 Moths and Butterflies (Lepidoptera)

These are insects under the order of Lepidoptera. This is the most widespread ad widely recognizable insect orders in the world. Insects under this order tend to have many variations of the basic body structures that evolved to gain an advantage in lifestyle and distribution. They are usually characteristic of three derived features. The most visible being the presence of scales that cover the bodies, wings and proboscis. The scales are modified and flattened giving moths and butterflies a unique color and patterns. Most of them have a membranous wing. The moths and butterflies undergo complete metamorphosis.

The mating and laying of eggs are done by adults. It is usually conducted on the host of plants in the case of the larvae. They undergo complete metamorphosis. While in the larval stage, they are commonly referred to as caterpillars. While in the larval stage, the caterpillars are usually relatively different from the adult moth butterflies and moths. While in this stage they have a cylindrical body with a well-developed head, three pairs of thoracic legs, mandible mouthparts. They also have pairs of prolegs. As they develop the larvae change in appearance. They change in their appearance and undergo a series of stages known as instars.

When fully matured and the larva develops into pupa. In some cases, the moths and butterflies spin silk case. In the case of butterfly pupa called chrysalis that is made of hard skin having no cocoon. Once the metamorphosis, and emerges as mature adults. There a variety of winged patterns and coloration and complex-patterns butterflies. They play a major role in the natural ecosystem and more so as pollinators and in the food chain. While in the larval stage they are considered to be very problematic to vegetation. In the case polymorphism is the appearance of forms such as the morphs that differ in color and number in attributes within the single species. In the case of the environmental polymorphism in which the traits that are not inherited. In life cycles, they composed of the egg, larval, pupa and adult. In this case of the pupae of the moths have been encapsulated in silk known as cocoons while in the uncovered pupae of butterflies that chrysalides.

While in the mating the males commonly start with the eclosion before the females. Both of the sexes tend to be sexually mature by the time of the eclosion. In normal cases, the butterflies and moths normally do not associate with each other. This is in exception of the migration species who tend to be social. The mating is initiated by the adults where they attract males. It is normally through the use of visual stimuli utilized by diurnal species like most of the butterflies. The moths and butterflies are an important part of the natural ecosystem as they are greatly involved in the food chains. They are co-evolved with flowering plants and predators hence the trophic relations.

1.5.15 Nerve-winged Insects (Neuroptera)

These insects are of the order called Neuroptera. Insects in this category are known to be six- legged. The name of the order is derived from the Greek word neuron which means cord and the other part ptera which means wings. In as much as insects from these groups are said to be nerve winged, the outstanding thing is that their wings are not laced sinews or nerves but branching veins and cross veins. Insects in this order vary in that they have been grouped into three distinct orders, Neuroptera being among them. In the case of the adult nerve-winged insects have two pairs of membranous wings almost equal in size having many veins.

Figure 1.14: Nerve-winged insects (Neuroptera).

[Source: http://lyndarimke.com/portfolio-item/lacewing/]

In the Neuroptera wings, they have abundant cross veins more so near the leading edge of the wings in between the costa and sub-costa. The insects under this order have chewing mouthparts and filiform antennae with many segments. It is important to know that the nerve-winged insects are weak fliers. Their larvae are elongated. They also have squared heads and long thoracic legs. In the case of the larvae of nerve-winged insects are predaceous and chewing mouthparts. The nerve-winged insects usually undergo complete metamorphosis. Their life cycle involves four stages involving egg, larva, pupa and adult. While in the Planipennia, they produce silk from the Malpighian tubes.

The silk is usually extruded from the anus and used in the spinning of the cocoon. Other nerve-winged insects usually have a naked pupa. Most of the insects under this order are terrestrial. The larvae in this case live in aquatic environments while the adults live near the water.

1.5.16 Parasitic Lice (Phthiraptera)

Lice are said to belong to the order Phthiraptera and tend to lean towards the parasitic group. These insects have a variety of hosts including birds and mammals. Their life cycles involve them being in a single host. The transmission is highly possible when the host comes into contact with other animals or during the mating process. This has required them to have a unique adaptation so that it can meet the ecological niche while on the host. The feeding habits vary in the lice as some of them feed only on blood, others feed on feathers while others on fur.

Some of the characteristics of the lice is that they are wingless ectoparasites, they have three nymphal instars, they have a simple metamorphosis, their mouthparts have been modified to allow for chewing and in some other insects, they are used in sucking. They have tarsi made of one or two- segmented, they are dorsoventrally flattened head, the operculum development occurs in the egg. They have reduced labial palps. They also have reduced compound eyes with no ocelli while their antennae are 3 to 5 segments. This order has been divided into two which include Mallophaga which is of the chewing lice and the Anoplura which is of the sucking lice. The chewing lice are the one with the largest population.

1.5.17 Rock Crawlers (Grylloblattodea)

These are insects under the order of Grylloblattodea. They are tiny wingless insects. Their heads resemble one of the cockroaches. They have long antennae as well as elongated cerci that arise from the tip of their abdomen. They cannot tolerate warmth. Though the long cerci, they don't have wings. Their eyes are reduced and have no ocelli. Their closest relatives being the Mantophasmatodea. In the case of these species, they have restricted distributed and small populations. The increase in global warming makes the species to be endangered. These insects are usually omnivorous though they feed on dead arthropods and carrion. When there is a reduction in the amount of carcasses their diet relies on the plant materials.

1.5.18 Scorpion Flies (Mecoptera)

Insects in this category belong in the order of the Mecoptera and the scorpionflies are in the largest family. In these insects, the males have an enlarged genital which looks similar to stingers of scorpions. They bear a somewhat fly-like appearance, them being small to medium -sized insects with long slender bodies. They also have narrow membranous. The breeding among these insects is conducted in a moist environment which includes leaf litter or moss. The eggs need the wet seasons to allow them to hatch. The larvae are usually in a caterpillar-like and are known to feed on vegetable matter. The non-feeding pupae may go through a diapause until the weather conditions are favorable.

Figure 1.15: Scorpion flies (Mecoptera).

[Source: https://edu.glogster.com/glog/scorpion-fly-46290185/27830dtnv50?=glogpedia-source]

The adults are known to be consumers of dead organisms and are the first insects to arrive at a cadaver allowing them to be useful in forensic entomology. These species occur in diverse areas but there is greater diversity in the generic and family. These insects have great importance when it comes to evolution. These insects have simple mouthparts and a long labium, long mandibles, and fleshy palps that resemble those of primitive true flies. Like other insects, they possess compound eyes located on the sides of their heads having three ocelli on the top. They have antennae that are filiform and contain multiple segments.

The fore and hind wings bear a resembling each other. They are long and narrow having numerous cross veins that are similar to those of primitive insects. The abdomen is cylindrical having eleven segments the first one of which has been fused to the metathorax. There are the cerci which have one or two segments. The abdomen curves upwards in the male there is a tip that contains an enlarged structure known as the genital bulb. They have caterpillar-like larvae that have a hard sclerotinized heads that have mandibles, short true legs on the thorax, there are prolegs found on the eight abdominal segments. There are suction disc or pair of hooks present on the terminal tenth segment.

The pupae are said to have free appendages compared to them being secure within a cocoon. The scorpionflies are known to live in plentiful damp leaf litter. In the mating process, the males initiate the process by emitting pheromones that attract mates.

1.5.19 Silverfish (Thysanura)

This is a small insect wingless insect. This insect bears a fish like a resemblance when its movement and has a silvery light grey color. The diet of this insect is composed of carbohydrates including starches and sugars. These insects have abdomens that tamper at the end hence the fish-like appearance. When newly hatched they are whitish but later on the develop a greyish hue and metallic shine as they progress in age. They tend to have two long cerci and one terminal filament in the tip of the abdomen that is present between the cerci.

Figure 1.16: Silverfish (Thysanura).

[Source:https://wiki.bugwood.org/NPIPM:Thysanura]

They have two compound eyes even though the other members of the zygentoma are completely eye less. Silverfish are wingless. They have long antennae and move in a wiggling motion which is similar to that of a fish. Also, the appearance of the silver color and scales on its body bring about the common name. These insects live about two to eight. They are also agile runners and have the ability to outrun most of their predators but this kind of locomotion is only possible on horizontal surfaces. They do not have additional appendages. This implying that they are not fast enough when climbing walls at the same speed. These insects are not friendly to light.

In the reproduction stage, it occurs in three phases. The first face involves the male standing face to face with the females. Their antennae touch then the back off and the process is repeated. Then return to the face to face position. While, in the second phase involves the male running away while the female chases him. The third phase involves the male and female standing side by side and head to tail while the male vibrates its tail. The male then lays the spermatophore which is a sperm capsule that is covered in gossamer and the female takes it to her body through the ovipositor so that they can fertilize her eggs. The eggs are usually whitish and are oval -shaped and make take between two weeks to two months to hatch.

The females lay fewer than 100 eggs in a lifetime. Once the nymphs hatch, they become whitish color and resemble smaller adults. They then undergo a molting process where they become greyish in appearance and have a metallic shine. They then become adults. These insects are unique as the molting process continues even after they become adults. They are also considered to be the most primitive insects. The silverfish feed on the matter having polysaccharides including starches and dextrin. They are known to damage the wallpapers so that they can consume the paste. They are also known to cause damage to the tapestries. They also feed on cotton, insects, linen, silk and also leftover crumbs. They are known to live up to a year without food if the water is available. To some individuals, they are considered to be household pests.

2

Phylogeny and Evolution of Insects

CONTENTS

2.1 INTRODUCTION

Creatures have been in existence for ages since the beginning of the world. As the ages advanced, features of different creatures had to change for better adaptability to their environment. The changes in their structural features or behavioral features have been caused by the alternation of the characteristics of their surroundings due to different factors such as climate, their population among others.

The changes in both the environment and different creatures do introduce us to the terms evolution and phylogeny.

Evolution is the slow but not sudden change in the characteristics, the ones passed from parent to child, of a certain species in its generation.

Phylogeny is the study of the different groupings of the many creatures which have been in existence for ages. The groupings are basically because of their ever -changing surroundings.

Insects are the small creatures that always surround human beings in their environments. To be more specific, based on the scientific meaning of insects, they do refer to the six-legged creatures whose body is usually divided into three parts and some of them might have wings while others might lack wings based on their phylogeny.

Evolution has many stories of how it was first discovered. All the stories do come to the same conclusion. They describe that certain species of creatures did originate from a single organism. Further explanations do outline that, as the offsprings of that single creature did continue to reproduce, they had to distribute to different environments. The movements are said to have been caused by their population increase forcing them to move to surroundings with enough space to accommodate them (Rasnitsyn & Quicke, 2007).

The new surroundings did have different characteristics in terms of temperature among others which forced them to find ways to adapt to the changes. Some of the species found it hard to persevere the change and did end up dying. Others did survive because of the theory commonly referred to as survival for the fittest. The survival of the fittest theory does state that only the species whose body structure was healthy could survive in their new environments.

Species that did survive were also said to have been naturally selected by nature, that is, many believed and do believe to date that nature has a way of selecting its species.

As the species were selected, they had to change some of their characteristics for them to completely fit in their new environments. As they continued to change their features, so did their environment because of many factors such as climate, increase in the population of the species among others. All this meant that the species would forever keep on changing, hence, evolution was discovered. With each new generation, new features were discovered in them. The evolved species did have a common characteristic with their previous generation only that some of their features were more advanced for them to perfectly survive in their new environment (Rasnitsyn & Quicke, 2007).

2.2 HISTORY OF PHYLOGENY

The term phylogeny is a Greek word that means race origin. After a thorough study of the origin of a certain species, evolutionists thought of a way to represent their information in a simpler way for easy understanding. Phylogeny was introduced to help in categorizing species into different groups. The groupings were later scrutinized thoroughly for a better understanding of their similarities and differences. For a better understanding of what they represented, the phylogeny strategy was later advanced such that the groupings would be represented in a tree structure or hierarchical organization, that is, from the most superior species, the origin of the rest, to the least superior species (Rasnitsyn & Quicke, 2007).

Insects have been in existence for ages. Most of the articles claim that insects date back four hundred to five hundred million years ago. To be more precise, insects existed from the word go when the universe was created. Their original structure back in the days has changed though there are some unique characteristics in them that seem to never change. As discussed earlier, they had to change for them to adapt to their environments which kept on changing with each passing day.

Originally, insects were known to be six-legged creatures with only three body parts. Some have retained their original structure while others have evolved to be a bit different such that they can fly as others crawl.

Insects are largely classified as a living organism which has been able to survive for ages in the history of the earth. They are part of the classification commonly referred to as Kingdom Animalia. The Animalia kingdom is further divided into phylum, which is its smaller classifications. Insects are categorically in the phylum of arthropods. The phylum of arthropods is further divided into classes and one of the classes is the class of Insecta where insects are known to belong to (Dudley, 2002).

Back in the ages, insects are said to have had same features be it in terms of body structure, behaviors among others. As the history of the earth kept on changing, so did their features for they had to find a way to survive and fit in.

In the struggle of trying to fit in, some insects did end up being instinct while the others which were strong enough did survive. The surviving ones developed stronger features to strengthen them in their continued fight for survival. Some did end up developing the flight mechanism for from the word go, insects are believed to have lacked the power of flight. Most of the insects known to date are part of the generation that did survive the adversities of the history of the earth. The current insects are still in the battle for their lives for the history of the earth is still changing with each passing day (Engels, 2012).

Evolution has produced a number of insects we see today. The evolutionary connection of bugs to other animal sets remains unclear. Even though commonly grouped with millipedes or centipedes—probably in the context of convergent adapting to terrestrialisation. Evidence has furthermore materialized favoring closer evolutionary connections with crustaceans.

Insects form one clade, closely associated with myriapods and crustaceans. Different other terrestrial arthropods, like centipedes, scorpions, and spiders, furthermore are occasionally confused with bugs since their bodies can look similar, having in common a connected exoskeleton. Nevertheless, upon closer inspection, their features vary considerably; most remarkably, they don't have 6-leg characteristics of adult bugs (Capinera, 2011).

The greater-level phylogeny of arthropods remains to be an issue of debate and investigation. In the year 2008, scientists discovered the world's oldest recognized whole-body impression of a primeval flying bug, a 300-million-year-old sample from the Carboniferous era. The oldest distinctive bug fossil is called Devonian *Rhyniognatha hirsti*, which originates from the 396-million year Rhynie chert. This might have superficially looked like a present-day silverfish insect. The species already contained dicondylic mandibles (a pair of articulations within the mandible), a function related to winged insects, showing that wings might already have developed at this time. Therefore, the original insects possibly appeared earlier during the Silurian era.

There are 4 major super evolutions of bugs that have occurred in history: Beginning with beetles (roughly 300 million years past), flies (around 250 million years past), moths plus wasps (roughly 150 million years back). These 4 groups account for most of the identified insect species. Moreover, some of the most successful insect groups today evolved in concurrence with flowering plants, showing some form of coevolution between the two species (Capinera, 2011).

2.3 THE ORIGIN OF THE POWER OF FLIGHT IN INSECTS

The origin of flight in insects still remains to be a mystery to many. Theories have been invented to try and explain why insects developed the power of flight.

The theory of water-skimming does have its explanation. It states that terrestrial insects developed their power flight as an advancement of their skimming capability on water. The original terrestrial insects are said to have developed the habit of frequently skimming on the water surface for fresh air. With the change of earth history, many of the terrestrial insects had to develop wings for them to fit in their new surroundings because of factors such as the attempt to save their lives from their predators (Price, 2003).

The second theory is that, once the first forest was developed, insects used to glide to access their prey from the canopies of the forest trees. As their environment kept on changing, access to their prey seemed to be a hard task for them. They had to develop better structures for them to perfectly survive in such environment hence some developed wings so as to be able to fly around searching for food.

The other basic explanation is that some insects modified the muscles attached to their exoskeleton to be wings for them to be able to fly. Such modifications are said to have occurred as the evolution of insects advances from one generation to another for it is not clearly stated when exactly insects developed their power to fly (Price, 2003).

2.4 INSECT FOSSILS

Insect's fossils are basically the remains of the earliest generation of insect species. They are mostly collected by scientists as they try to understand the origin of insects and how they have evolved over time. It is really hard to collect insects' fossils since insects are minute creatures and the discovery of their remains is almost impossible.

Fossils are later preserved for future reference in situations where individuals need to further polish their knowledge on insects.

2.4.1 Preservation of Insects' Fossils

The fossils are preserved in different ways. Preservation methods do vary depending on the survival environment of that species. This is because they have to be preserved using techniques that ensure that they are in good condition. Their preservation methods are generally not complex since they do have a hard exoskeleton unlike the preservation of the soft-bodied creatures.

Preservation of freshwater and marine insects' fossils: Insects survive in different environments. Once they die, their remains might remain in their same habitats or carry to other habitats. Insects whose remains are carried by water or wind to the watery environments may still be able to survive. Their survival depends on whether the animals living in watery environments are ready to use them as their prey or not. The insects remain which survive are preserved in the water body's bottom levels. The bottom levels do maintain the fossils in perfect conditions since their general characteristics are quite suitable for the survival of even dead organisms (Shaw, 2014).

Once the fossils once preserved from water bodies are collected, they need to be re-preserved in environments whose characteristics match the ones found beneath the water body's surface, be it temperature, salinity percentage, chemical characteristics among others. All these conditions must be met in the new preservation environment for the remains to still be intact.

Preservation of insects' fossils by compressions, impressions and mineralization techniques: Insect remains may undergo compressions, that is, maybe squeezed and reduced in size due to many factors, based on the environment in which they have remained since the death of that insects. For insects that die on rock surfaces, or whose remains are relocated to rocky areas either by wind or any other factor, they do undergo compression. Compressed fossils are later re-examined for the separation of the insect fossils from the new mineral contents which have formed around the fossils over the years through the mineralization process. Mineralization occurs when more and more mineral content either from the current surrounding or ones brought from new environments accumulate around a certain species. After the accumulation process is complete, a new impression, that is, a different design and structure of the original fossil remain is formed (Shaw, 2014).

For archeologists who deal with the preservation of insects' fossils, they have to do lots of extraction of the minerals which have accumulated around the fossil using various mineralization techniques based on the type of mineral they discover to be present during the laboratory examination of the fossil. Once the extraction process is complete, the discovered fossils are preserved under suitable environments which should match the chemical characteristics of the mineral contents found to surround the insect fossil at first. Such specifications ensure that the insect fossil remains to be in perfect condition for use in future in case of any references needed to be done on them or for later further study.

2.5 THE EVOLUTIONARY THEORY OF INSECTS

Insects are believed to have existed for ages. Some have been extinct as new species continue to be discovered due to environmental factors which keep on changing. General evolution theories of insects are based on the forced adaptation techniques which insects have to experience as they come across the pressure of fitting in a certain environment for their survival (Engels, 2012).

Many believe that the evolution of insects is mostly dependent on the thriving of flowering plants for most insects depend on flowering plants. A bigger percentage of plants depend on insects for pollination of their species, hence the more the plants to be pollinated, the more the number of insect species is needed to exist to help in a balanced symbiotic relationship between the plants and the insects. The symbiotic relationship ensures that insects can get food for their survival from the flower nectar of the plant while the plants get pollinated for its reproduction process to take place. All of this does ensure a win-win situation.

2.5.1 The Devonian Evolutionary Theory

The earliest insects are believed to have been formed during this period. The Devonian period is said to have been a period of high concentration levels of carbon dioxide in the atmosphere. The high concentration levels led to high-temperature rates which facilitated the evolution of the first insects (Rasnitsyn & Quicke, 2007).

These first species of insects are said to have developed more advanced and better tracheal systems. The advanced tracheal systems ensured that the insects did survive in environments of low oxygen levels.

2.5.2 The Carboniferous Evolutionary Theory

Carboniferous is a term used to refer to the period when components that were made up of carbon components and other chemical components whose characteristics are close to that one of carbon accumulated in one place. The accumulation process is believed to have taken place a long time ago. The forests which existed back then were affected by the accumulation process for most of them were destroyed by the deposits which accumulated most of the space around their surroundings leaving them with no enough space to acquire adequate amounts of oxygen and carbon dioxide. Insects that resided in those forests were buried in the deposits and left to die for there were no rescue means for them (Dudley, 2002).

It is believed that most of the insects which did exist in that period had the power of flight. Archeologists who specialize in the preservation of insect fossils did find remains of insects as they excavated on the carboniferous. The fossils are supposed to be preserved in conditions that match the ones found in the archeological location. This is because the carboniferous surroundings did have special chemical components which helped in the maintenance of the fossils (Dudley, 2002).

2.5.3 The Permian Evolutionary Theory

The Permian evolutionary theory is a period in which all the earth's land masses accumulated in one place during the history of the earth. The landmasses collapsed from highland regions and at the end accumulated on the lowland level of the earth. As the masses accumulated, most of the insects which did reside on the ground surface of the earth were all buried. Later, during the excavation process, their remains were discovered and preserved in perfect conditions for future references. Insect fossils discovered during this period are commonly referred to as the Permian insect group (Price, 2003).

2.5.4 Triassic Evolutionary Theory

The Triassic is known to be a period when arid and semiarid savannas did come into existence. Insects from the previous periods of the earth's history did evolve for them to perfectly survive in their new environment. Most of the insect fossils from this period have not been discovered since the climatic conditions of those days were not suitable for any preservation purposes. However, some remains might have been transferred to areas with better preservation conditions and later discovered by archeologists and preserved under better conditions for future references.

2.5.5 The Jurassic Theory of Evolution

The Jurassic period is commonly known for it brought about the existence of birds. Birds to date happen to be the number one predators for insects. The recovery of insect fossils was a hard task since most of their remains back then were used as prey by birds. The climatic conditions of the Jurassic period were similar to those of the Triassic period hence the perfect preservation conditions of any insects remain was not that much possible. Archeologists were a bit lucky in that some insect remains were carried away into other locations by agents such as water or wind. Their new locations did seem to have better preservation conditions hence they did remain in perfect shapes. Once the insect fossils have been collected, they are preserved under perfect conditions for their future use (Price, 2003).

2.5.6 The Cretaceous Evolutionary Theory

The Cretaceous period is known in the earth's history to be the period when, the large combined mass of the earth's surface, which was referred to as Pangea, split into different portions forming the known continents of the earth (Price, 2003).

With the splitting of the earth's surface, some portions of land did submerge as more water bodies were being formed. Some of the insects from the previous period of the earth did become extinct. Other insects had to undergo yet another stage of evolution for they had to find a way to fit in their new environments. Climate conditions of the different portions of the earth's surface did change due to their new locations, different from their original locations in which almost all the earth had the same climatic conditions.

Some insect fossils have been collected from different parts of the earth's continents though they do have different characteristics from their previous generations due to the evolution process which took place under the forced circumstances which the insects faced. They are preserved under perfect conditions for future references (Price, 2003).

2.5.7 The Paleogene Evolutionary Theory

The Paleogene deals with the period when some of the earth's surface was covered with clay deposits. The deposits happened to contain insect fossils. Archaeologists interested in the study of insect fossils did excavate them from the clay deposits. The fossils were later preserved under perfect conditions for future use such as for study purposes (Price, 2003).

2.5.8 The Neogene Evolutionary Theory

Neogene is commonly referred to be a period when all insects on the different continents of the earth had fully adapted to their new environments and were thriving perfectly with little disruptions from the ever -changing climatic conditions. This is because after the splitting of the earth into different portions, the climatic conditions did stabilize for a while and in case any change occurred, it was quite slow for it to have had immediate effects (Price, 2003).

The insect fossils were collected and preserved. Most of the fossils collected in this period are of many similarities to the modern generation of insects known to us.

2.5.9 The General Structure of Insects

For something to be said to be whole, it has to be made up of various subsystems or parts. This does apply to insects. Insects are made up of different components that ensure that they carry out their day to day lives in the most effective way. Some of the parts can be visible from their outside structure while others can only be visible if the insect is dissected. The study of the structure of insects by human beings is referred to as insect morphology (Price, 2003).

When it comes to insects, they are known to be under the classification of phylum anthropoids. Insects are said to be the class under phylum anthropoids with large numbers. Insects are distinguished from the other classes in phylum anthropoids in that they have their bodies are divided into three different parts (the head, thorax, and the abdomen), have three pairs of legs and their mouth part is outside t head capsule.

Insects are known to be of various species which vary in terms of their body structure. Some are known to lack eyes, others have eyes, others have legs which have been modified for running, jumping, swimming and some digging. All the modifications in the various insect species give them a chance to fit in almost every ecological niche of our planet earth except the deep oceans and the extremely cold surroundings.

The insect body is made up of a hard outer layer known as the exoskeleton which protects the entire body of an insect. As stated earlier, the insect is made up of three body parts. The head is known to specialize in sensory functions which enable the insect to better understand its surrounding and it's also used for food intake. The thorax which is the middle body part, is known as the pivot of the insect. It does hold the locomotive structures of the insect such as wings and legs. The abdomen is the last body part of the insect in its rear end. It is known to specialize in respiration, reproduction, digestion of food, and excretion of waste substances from the internal body of an insect (Price, 2003).

2.6 INSECT PHYLOGENY

We have stated on several occasions that with time, insects have evolved into different species depending on their geographical location. The changes have been there to enable different insects to adapt to their new environment for survival purposes. With the change in the earth's history, so has the different geographical regions changed. This change has been a major factor that has contributed to the continued evolution of insects into different and more species. This is evident because, with each passing day, we have new insect species being discovered.

All the discovered insect species have been classified into different groups for a better understanding of their characteristics, similarities between different species and their differences. Insects have been known to belong to a kingdom of living organisms known as kingdom Animalia since they are animals for they are able to move, eat, reproduce, and excrete, among other characteristics which are only found in animals. The kingdom has been further classified into different phyla. Insects belong to the phylum of

Antropoda since they are made up of an exoskeleton on their outer surface. The Antropoda phylum has further being divided into different classes with the Insecta class being the class into which insects belong to. Class Insecta is also known for its large numbers compared to the other classes in phylum Anthropoda. From the Insecta class, insects are further classified depending on the order, then their families, followed by their genus and later their species. All these classifications are distinguished by the different features in the many species of insects for a better understanding of them (Shaw, 2014).

After classification, the different groups which have been formed are given a unique name that has to follow the rules of binomial nomenclature. Binomial nomenclature rules state that the insect groupings need to be named using a Latin name for the Latin language does not change over the ages; the name should be written in italics if types and if handwritten it should be in underlined; the genus name should be separated from the species name; among other rules.

One of the phylogenies of insects is the Apterygota phylogeny. It is made up of insects that do not have wings. We did state earlier that originally, all insects were wingless, but as the evolution process continued, a larger percentage of insects developed wings. The remaining wingless insects of our current age are the ones referred to as the Apterygota. The Apterygota insects are further subdivided into different groupings which are the families according to classification rules (Shaw, 2014).

One of the families is the Archaeognatha which is also referred to as the microcoryphia. They are known to be very active in nature in that they ever alert and ever -moving or carrying out their daily activities. Archaeognatha are cylindrical in shape with their length measuring not more than twenty millimeters with their antennae being quite long for better sensory mechanism. Their mouthparts are not that visible unless with careful scrutiny or by use of evolutionary studies. The body is covered with slightly hardened scales which help in the protection of the internal body structures. The other family of the Apterygota is the Zygentoma. They are said to have similar characteristics with the Archaeognatha family with their only difference being in terms of their mandibles. Zygentoma mandibles are more of dicondylous in nature. They are said to have been discovered almost the same time with the Archaeognatha family and that is why they are so similar to them but with further scrutiny, the evolutionists were able to identify their distinguishing factor which is the mandibles (Snodgrass, 2018).

The other order of class Insecta is the Palaeoptera. Palaeoptera is an is an order which consists of insects that have wings but cannot fold their wings with the help of the fold lines and flexion lines over their abdomen. They lack their fold lines and flexion lines in their abdominal wing section. The Paleoptera order is further divided into different families for its better understanding.

One of the families is the Ephemeroptera family. This family is known to comprise of the most primitive winged insect family. Their wings are quite unique in nature for they are triangular in nature more for their forewings if held vertically above the rest of the body. The adult species of the Ephemeroptera family is known to reside mostly in the freshwater bodies which provide them with food and a breeding site for their reproduction activities. The other family of the Palaeoptera order is the Odonata family. The Odonata family is said to have a completely different appearance from the ephemeroptera family even though they belong to the same order. The wings of the Odonata family are long, narrow, and parallel to each other for both the forewing and hindwing. The forewing and the hindwing are known to be of the same measurements, be it in terms of length or width, they are the same.

Polyneoptera is also an order of the class Insecta. This order is known to be one among the higher groups of winged insects. Their wings have well developed and perfectly functioning fold lines and flexion lines since the wings can fold for over the body. The folding mechanism gives them the advantage of being able to protect the entire wing from any kind of damage as they crawl through vegetation or any other confined place in search of a meal. With time, the Polyneoptera order has been able to modify the structure of its fore wings by making them be thicker to act as a protective cover for the hindwings and the abdomen body part of the insects. This order is known to be one among the many orders which have been able to modify their structures to fit in their ever -changing ecological niche. Just like any other order, the Polyneoptera have been further classified into different families (Snodgrass, 2018).

Embioptera is one of the many families of the Polyneoptera order and it is also referred to as the Embiidina family. This family is further grouped into several species which are said to be approximately four hundred and fifty in number, the known species so far. The Embioptera family is known to thrive perfectly well in tropical regions of the globe. The web spinner insects are one of the many examples of the Embioptera family. The

Embioptera family is said to be a sister family to the Zoraptera family which is also kind of related to the Plecoptera family that has similar characteristics with both the Phasmida and Orthoptera families. This connection shows that the Polyneoptera order comprises of more families as compared to the other orders of the class Insecta.

The second family of the Polyneoptera orders is the Grylloblattaria family which is also referred to as the Grylloblattadea family. It is more of a small family that is made of a few species that are said to be approximately thirty in number. This family's species are known to be dominant in high altitude regions which provide them with perfect climatic conditions for their survival. One of the examples species of the Grylloblattadea family is the rock crawlers insects found in high altitude rocks or the ice crawlers found in regions known to have glaciers (Javed & Agurla, 2016).

Mantophasmatodea is also a family in the Polyneoptera order. Thus the family is said to be one of the most recently known insect family which was discovered in the early months of the beginning of the new millennial period. Since it is one of the recent insect discovery, the species linked to it are quite a few in number for they are said to be approximately twenty species which have been associated with this family. Thus the family is said to be dominant in the African countries more so in the southern and eastern regions of the Africa continent. It is said to have strong links with the Grylloblattadea family in terms of its characteristics in general. The heelwalkers insect is one of the species that belong to the Mantophasmatodea family. The name heelwalkers is said to have been derived from the fact that the end of the tarsus leg segment is always raised above the ground as they walk around from one region to another in search of a meal, shelter or something else.

Orthoptera is a family among the families Polyneoptera order. It is said to be the only family with the most number of species for so far, the discovered species seem to be uncountable and with each passing day, more and more species of this family are being discovered all over the world. Some of the examples of its species include locusts, grasshoppers, crickets, and bush-crickets all of which are known to be destructive pests to food crops. The fact that the locust is well known to many shows that the orthoptera family has been well studied and its impact on human life is known.

Phasmida which is also known as the Phasmatodea or the Phasmatoptera family is also part of the Polyneoptera order. They are known to many because of their unique feature of mimicking the color of tree twigs or leaves once they attach to them. The eggs of this family insects are said to resemble most plant seeds. An example of this family insect is the stick-insects.

Plecoptera happens to be one of the families in the Polyneoptera order. Their nymph system is said to always be aquatic in nature a characteristic that differentiates them from the rest of the terrestrial nymphs. Stoneflies are one of the species which belong to this family (Rasnitsyn & Quicke, 2007).

The Dermaptera family is part of the large Polyneoptera order. They are said to have developed forceps-like cerci which make them unique among most insects. Dermaptera family is dominant in the cool regions of the globe which provide them with the proper climatic conditions for their survival. An example of this family species is the earwig insect.

The other family of the large Polyneoptera order is the Dictyoptera family which is made up of few species that include the cockroaches, termites, and mantids.

The last family of the mighty Polyneoptera order is the Zoraptera family. It is said to have at least thirty-five different species that have similar characteristics but do also differ in one way or the other. They are dominant in tropical regions for they are known to reside in rotting woods because they feed in fungal hyphae. They live in small groups commonly known as colonies to many.

The Palaeoptera order is the fourth order of the class Insecta. If is also made up of several families that vary in terms of their general structure.

One of the Psocoptera family whose species include the booklice and barklice.

The other family us the Thysanoptera family which are said to be quite tiny a characteristic which makes them be less known to many. Some feed on fungi, others on higher plants, and a small percentage of the species in this family tend to be a part of the agricultural pests. An example of this family's species is an insect referred to as a thrips.

Hemiptera is a family that is known to many because of their elongated mouthparts and also because it hosts the species of the true bugs insects.

The last family in the Palaeoptera order is the Pgthiraptera family. They are not that known to many. An example of their species is the sucking and biting lice.

The last commonly known order of the class Insecta is the Endopterygota order. Its larvae are said to be completely different from its adult body for it undergoes metamorphosis in a pupa form with its wings developed internally. Just like the other orders in the class Insecta, the Endopterygota order is also made up of different families.

One of the families is the Coleoptera family whose species tend to survive in multiple habitats. Their species include insects such as the beetle whose appearance is quite unique since most of their outer body is covered in a hard layer which melts down towards the mid-life of the insect outer body.

Diptera family compromise of the true flies insects.

Hymenoptera family does comprise of the species such as the ant, bees, wasps among others.

The Lepidoptera family comprises of the butterflies and moths species among others. They have flattened scales on their wings which makes then quite different from the other families.

The Trichoptera family comprises of the caddisflies and the sedge flies.

The Strepsiptera family is one among the strangest families of the Endopterygota order. One of the many species belonging to this family includes the Stylops family.

Siphonaptera family does comprise of basically all the flies families.

Raphidioptera family is known to resemble the shape of a snake in terms of their general structure hence most of their species are referred to as snake flies.

Neuroptera family comprises of lacewings species among others.

Megaloptera family is made up of the alderflies whose adults tend to have a flattened head with their biting mouthparts quite visible from afar with two pairs of wings attached to their bodies. The larvae of this species are known to thrive well in freshwater surroundings.

The last family of the Endopterygota order is the Mecoptera family which comprises if scorpionflies species.

2.7 IMPACT OF INSECTS TO SOCIETY

Insects are known to be everywhere around the globe. From the farms to our shelters, to our beds, basically all over our surroundings. By being all over, they have affected the livelihood of all the other living creatures. They are known to destroy agricultural produces, they are known to destroy the beautiful shelters which have been constructed by human beings, they are known to cause all sorts of illnesses to both human beings and other animals, they do provide human beings and animals with food, promote decay, damage stored grains among other effects.

In terms of economic effects, insects are known to cause large losses. The crops which are destroyed by insects would have been used to provide food for many which means governments have to purchase food for their people. The control and prevention of paste invention to food produce is also quite costly since farmers are supposed to purchase pesticides or insecticides that tend to be quite expensive for the average people in the society. The insects which transmit diseases to human beings force them to incur a cost for their treatment. Insects have affected the economy of many parts of the globe negatively.

The insects might be known for the negative effects they cause, but they do also have positive impacts. They help bury the dead since they fasten the decaying process of organic substances. The same insects which are hated for destroying crops help in promoting soil fertility. This is because they contribute a lot to the decomposition of organic materials which add more nutrients to the soil. Insects promote the soil aeration. The boring insects are known to create small passages in the soil which provide facilitate the movement of air to and from the soil more so the deep layers of the soil. Insects do have a special way of maintaining the natural environment in its normal healthy state of being ever clean and productive. Some cultures of different communities do consider insects to be sacred in a way hence insects have promoted different cultures that identify us in the society in general. The evil caused by insects is somehow balanced by the good caused by the same insects.

Insects do provide several living creatures with food. Animals such as birds depend on insects as their source of food. Fish species feed on aquatic insects. Some human beings, do collect different insect species and prepare their meals. The human habit of feeding on insects is known to have originated from our ancestral roots for our ancestors did feed on insects.

In terms of agriculture, insects help in aeration of the soil as stated earlier. Beetles help in the distribution of waste materials from big animals such as the dug of cows which boosts the soil fertility. Insects help in improving the water retention of the soil. They do pollinate the farm crops which promote their production. They feed on other insects which tend to be quite destructive in our farms hence reducing the number of predators that destroy our farm produce.

Insects are known to provide us with useful products which boost society. Bees produce honey which is known to have great health benefits to the human body. Beekeepers may earn a source of living by selling the products produced by bees. The wax found in beehives can be used in the manufacture of different products which improve the economy in one way or the other. Silkworm is known to provide people with silk which can be used in making silk clothes.

3

Insect Morphology and Anatomy

CONTENTS

3.1 INTRODUCTION

To understand how all living organisms are related, biologists have classified them into different groups. These groups are specified on the features that animals share. Animals are then given specific names so that people the world over can communicate the names and about the animals no matter the language and dialect. This is done by using traditional Latin words. Animals belong to a number of different groups but the classification first starts with the animal kingdom.

All living organisms placed into different kingdoms and there are 5 to classified life on Earth. These are:

- Animals
- Plants
- Fungi
- Bacteria
- Protests

Insects or "insectum" in Latin (meaning to cut into) fall under the kingdom of Animalia. The animal kingdom is divided into 36 smaller groups known as a phylum. The majority of the animal kingdom fit under around nine phyla with insects ,in particular, all ranked under Arthropoda. An Anthropod essentially is an invertebrate with an external skeleton a segmented body and appendages which are leg like jointed attachments. They also fit under the subphylum Hexapoda. Insects are in fact the largest group within the Anthropod phylum with more than 1 million prescribed species and over 30 million still left to be identified. They are the most diverse group of animals on the planet and consists of over 90 per-cent of animal life formed on Earth. Insects are diverse and can be found in nearly all environments on Earth however very small portion resides in the oceans which is dominated by another Anthropod known as crustaceans.

3.2 THE GENERAL CHARACTERISTICS OF INSECT ANATOMY

According to Snodgrass (2018), insect morphology is concerned with the discipline of the physical form of insects. The terminology describing insects are similar to those used among anthropology (the study of arthropods) in general because of their shared evolutionary. However, three distinct physical features separate insects from other arthropods.

- They have a body divided into three distinct sections (the head that bears mouthparts eyes antennas and sensory input organs; thorax where the legs extend from; and the abdomen that holds the digestive and reproductive organs
- They have three pairs of attached to the three- segmented thorax that is sometimes evolutionarily modified to run jump dig or even swim)
- They have mouth parts located outside the exoskeleton head capsule. (This and the position of the mouthparts is probably the most distinctive feature from their closest relatives)
- They have no internal skeleton and instead have a hard outer exoskeleton made of a glucose- based polymer.

In studying morphology, the diversification of insects is enormous and they can range from very small (0.3mm) to the largest measuring over 60cm (*Titanus giganteus*). Furthermore, some insects have many eyes, and some with none at all there are also insects with highly developed wings while others have none as well. When studying insect evolution one must consider the power of descent with modification and this is what allows them to occupy almost all the ecological zones on earth (Price, 2003).

3.3 INSECT TAXONOMY

Taxonomy is the discipline pertaining to the principle of scientific classification and so by extension insect's taxonomy is the study of the scientific classification of insects as a class.

Discussed before, Hexapoda that include all insects fall under the phylum of antipodal and is divided into two subclasses;

- the primitive wingless insects are known as Apterygota
- And the winged or secondary wingless class of Pterygota

The subclass of Pterygota is divided into two classes depending on their Metamorphosis;

- Exopterygota: have a simple metamorphosis without a pupal stage
- Endoptreygota have a more complex metamorphosis that includes a pupal stage.

Metamorphosis in insects is the taxonomical differentiation between the different subclasses whereby Apterygota have no metamorphosis at all except for maybe size even from the larval stage, Exopterygota members are characterized by simple metamorphosis from egg to nymph to an adult where the majority of the change is in external appearance and feeding habits remain the same throughout the insect lifespan. Endoptreygota members have the most change during life with the eggs hatching into larvae that actively feed and may or may not have legs. Wings are not visible at this stage and after molting severally a pupa is formed. The pupa is an externally inactive stage of metamorphosis where the insect does not feed or move and is encapsulated by a cocoon of silk or soil. Internally however drastic changes take place and by the end of the people stayed a highly active winged adult emerges. The larva and adults' versions of West animals live in different places and use different food sources.

Identifying insects first begins by determining the correct order and good observational skills with the use of good hand lengths and a key to identify the insect orders (Snodgrass, 2018)

To have a comprehensive breakdown of the physiology and morphology of an insect we will begin by describing all of its external features first.

3.4 EXTERNAL FEATURES OF INSECTS

3.4.1 Exoskeleton

An insect's exoskeleton is comprised of two main layers;

- The epicuticle: This is a small layer that is waxy and water-resistant.
- The procuticle: This is the inner layer which is made of chitin (a hard long-chain polymer derived from glucose) and other tough protein molecules. This layer also produces thick armor that insects will need through biomineralization. Biomineralization is the process in which organisms produce minerals to stiffen up existing tissues.

The procuticle can be further divided into another two layers; the exocuticle and the endocuticle. In some insects, they may be a third layer in between the two known as the mezocuticle. The endocuticle tough and flexible with many crisscrossing layers of chitin. The exocuticle is thick, rigid and has a component known as sclerotin as well as more fibrous chitin.

Sclerotin is a class of protein molecules that when initially formed is soft but becomes a hard material due to its cross-linked formation through a process that is very similar to tanning, like the hides of animals to produce leather. The sclerotin is produced under an enzymatic environment that is further encrusted in calcium carbonate making it much more resilient and tough. Biochemically, sclerotin is variable with different species producing different proportions of different chemicals. In the soft-bodied insects, some adults but more especially in the larval stage the exocuticle is highly reduced (e.g. the caterpillar or honeybee larvae as shown below). The comparison is also easy to note with modified and matured exocuticle like in beetles (also shown below)

Figure 3.1: Cuticles of some armored beetles are particularly rigidly sclerotized.

[Source: https://en.wikipedia.org/wiki/Arthropod_cuticle]

Figure 3.2: Honeybee larvae have flexible but delicate unsclerotized cuticles.

[Source: https://en.wikipedia.org/wiki/Arthropod_cuticle]

For insects that have developed flight capability. You will find that their respective flight muscles and nerve impulses are attached to the inside of the exoskeleton. A single nerve impulse can contract these flight muscles several times making their wings beat a lot quicker than would be ordinarily possible.

Because cuticles are not cellular in nature they do not grow and therefore once established an animal that continues to grow beyond its size has to periodically shed these external parts in a process known as molting. During molting or ecdysis, the insect internally digests as much of the exocuticle material as it can and in an enzymatic process that uses a molting fluid, dissolving away and then digesting much of the older sections. The epidermis layer just below the exoskeleton has epithelial cells that allow the animal to fully assimilate as it absorbs material (Javed & Agurla, 2016).

This digested material is used to build a new exoskeleton and once the new cuticle layer has formed the insect may spit the remaining parts of the old reabsorbed material along lines of weakness of the newly formed integument. These parts eventually end up shedding as the exoskeleton evens itself out but some species carry them as camouflage or protection. Exuviae the name given to the four things that would be shared and it can be important to some biologists as it may help identify the species of the animal and even the sex. The insects that either swallowing liquid or gas to stretch this new integument into shape and it's usually soft and pale but hardens along just like the initial formation of the procuticle and the epicuticle. The newly-formed exoskeleton is usually a little bit soft and peeled begin to begin with and then a hardening and pigmentation process takes place that can last anywhere between several minutes to several days depending on the insect and the environment.

During the embryonic stage at the very beginning of the insect's life, epithelial cells will give rise to the external cuticle with an internal membrane that will provide muscular support and protection. As an insect develops it will be digested from the inside and sequenced into material for the new cuticle during molting. The hardened plates that are formed along the exoskeleton are called sclerotin and they are divided into four principal regions on the insect body namely; the Tergum (dorsal) Sternum and 2 pleura (laterals) (Javed & Agurla, 2016).

Figure 3.3: Larva of beetle, family Cerambycidae, showing sclerotized epicranium; rest of body hardly sclerotized.

[Source https://en.wikipedia.org/wiki/Insect_morphology]

3.4.2 The Head

Insects have a heart sclerotized exoskeleton head capsule known as an epicranium. The majority of its sensory organs are located here including the antenna or soulless eyes and eating organs. It is important to note that even with insects with that are none or weakly sclerotized, most insects, even at the larva stage tend to have sclerotized head capsules like the larva of a beetle shown below however there are some exceptions like the larvae of the cyclorrhapha who tend to have hardly any head capsule.

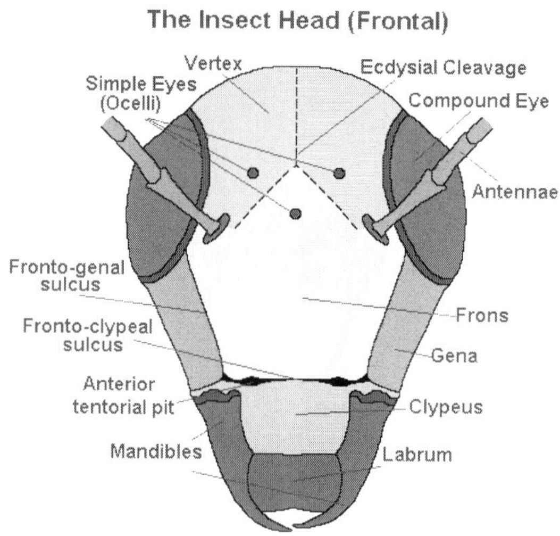

Figure 3.4: The head of an insect.

[Source: https://www.earthlife.net/insects/anat-head.html]

Orthoptera order of insects that consists of locusts, crickets and such display the greatest variety of features on the head of an insect including sutures that are too hard elements connected by a fairly rigid joint. In adults, the head is usually one piece but embryological studies show that the head of insects is divided into six sections holding a pair of appendages. The cranial Vertex or the upper surface of the head in insects can be commonly used to identify species especially using the Color and the Shape. This dorsal region is right in between compound eyes for insects that are hypognathous (have vertical downwards pointing mouthparts) and opisthognathous (receding mouthparts). In insects that don't have compound eyes but have ocelli, that are single lensed, simple optically arranged eyes, the Vertex would be located here.

They are sutures along the sclerites that make up the epicranium and they will be dependent on the insects moulting process. They can come in many different shapes either why Y, U or V and they split open providing an opening for the insects to emerge from the old integument.

Another important part to mention concerning an insect's head is the frons. The frons is located towards the anterior side or below the vertex where the antenna is usually located. It varies in size and its borders are arbitrary but generally, this area is the sclerites from which muscles related to and attached to the pharynx and mouth are located. The clypeus is another sclerite that separates the frons from cheeks (gena) and mandibles in primitive insects. The gena is poorly defined and falls right behind the compound eyes but articulations around this area and that of the mandibles can be highly sclerotized to support powerful jaws.

Connecting to the back of the head is a large aperture known as the foramen where organs such as the nerve chords esophagus and salivary ducts are connected from the head to the thorax. From the posterior angle, we can see occiput and the mandibles both labium and maxilla very clearly. Other details of the structure of an insect's head are shown below.

Evolutionary modifications of the insect head may occur and this is especially evident and species that have highly specialized sucking mouthparts. This is evident in the adult *Apis mellifera* and the larva *Heliothis zea*. The *Apis mellifera* for example has an epistomal suture that is significantly arched upward so much in a sense that it can be often incorrectly be identified as the frons. They have clearly identifiable posterior tentorial pits that together with other features are complex modifications to enhance the sucking mechanism.

3.4.3 Compound Eyes and Ocelli.

The majority of insects have one pair of very large compound eyes made up of very small units called ommatidia. A single compound eye can have up to 30000 ommatidia (e.g. large dragonflies) which give less resolution than what is found in vertebrates but provide a high-level of acute perception, especially of movement, along the ultraviolet spectrum with many having a sensitivity to green along the visual spectrum. The ommatidia in compound eyes are fixed along a convex surface and this point slightly in different directions providing a combination of different perception inputs. This also means that viewing angles are very large and insects with compound eyes have a lot better acuity than other animals. There may also exist an additional 2 or 3 ocelli to help adjustments in low light (Snodgrass, 2018).

Compound eyes are the visual organs among arthropods that can come in different types and are usually classified between either apposition eyes or superposition eyes. Apposition eyes form many inverted images that are combined in the brain and even if the ommatidia optically isolated. The more typical version of the apposition compound eyes exists in most day-active insects, like bees, where a lens focuses light from one direction on the rhabdom (optic rod) while the light from everywhere else is absorbed by the dark wall structured within the ommatidia.

The second type of compound eyes is the superposition eyes. Superposition eyes are further divided into three kinds:

- Reflecting: the reflecting and refracting superposition compound eyes of an insect are very similar however with reflecting apposition eyes there no gap between the lens and rhabdom is causing light to be directly reflected on the same axis that light enters the eye in.
- Refracting: the refracting superposition eyes have a small gap between the rhabdom and the lenses but have no side walls so each lens almost operates like its own eye taking in light at one angle and reflecting it at the same angle on the other side.
- Parabolic: Parabolic superposition compound eyes Focus light from a reflector or in this case a parabolic mirror in combination with a single lens. In a sense, it is the combination of both a reflective and reflective superposition eye. The lens will provide the focus but raise coming in from oblique angles will be intercepted and reflected by a parabolic mirror that covers the cones beneath the lens. This parabolic surface in the eye focuses and redirects light back across the axis of the ommatidia.

As insects are so diverse there are some that do not fit into any of the categories listed above and some exceptions include the so-called single lens compound eye which is something of a combination between a normal simple eye and a multi-compound eye. Another version is a pseudofaceted eye, which is a cluster of ocelli on each side of the head arranged to look similar to a compound eye. Some even have asymmetric compound eyes that involved in the *Temnothorax albipennis* ant scouts for example to continuously explore unknown nest sites while still having a view of the exits. All in all, the small size on an insect's lens, no matter which type of eyesight they have is limited in resolution. To put this in perspective a human would require compound eyes each the size of their heads to have comparatively the same resolution to our simple eyes (Snodgrass, 2018).

3.4.4 Antenna

Antenna or feelers are flexible attachments most often connected to the Vertex and are used for sensing the environment around them. Primarily antennas are used for the sense of smell in insects and there are fine little hairs known setae that cover antennas and allow them not only to smell and taste but in some species even to sense temperature, pressure, and humidity among a myriad of other useful sensory inputs. Antennae differ in size, shape, and segmentation across the insect order. The base of the antenna that is attached to the head is normally called the scape and the second segment is referred to as the pedicel. The subsequent sections of the antennae are known as flagellum the platinum can be further built up of many tiny units called flagellomeres (Steinmann & Zombori, 2012).

All insects have antennae and they develop from the torulus which is a raised portion on the epicranium. This area is built up just like much of the exoskeleton with sclerotized material and as much as they appear to hang loosely they are firmly pivoted and in many insects may begin to grow even from the embryonic stage. The scape and by extension the pedicel is attached to numerous muscles that allow the insect to move the antennae. Some insects actually have muscles running up through the entire flagellum while others have a membrane that is controlled by muscular movement (Steinmann & Zombori, 2012).

The flagellum may have attachments that are not true flagellomeres known as annuli that are small flexible interconnected rings. There are 13 different kinds of insect antennae shapes and forms listed below.

Table 3.1: Presents all the 30 orders of insects known at present. (Adapted from *Gullan P.J. & P. S. Cranston. 2010. The Insects: An Outline of Entomology*). It also provides information about the type of metamorphosis, wings and mouth-parts of adults

Order	*Examples*	*Wings*	*Type of mouthparts*	*Meta-morpho-sis*
Archaeognatha	*Petrobius maritimus*	None	Chewing	None
Blattodea	Cockroaches	Four	Chewing	Incom-plete
Coleoptera	Ladybugs, scarab beetles, lightning bugs	Usually four; hard, rib-less front wings and folded, membra-nous back wings	Chewing	Complete
Dermaptera	Earwigs	None or four; short, thick front wings, folded back wings	Chewing	Incom-plete

Diptera	Flies and mosquitoes	Two, very rarely none, back legs modified into halters	Sucking-licking or piercing-sucking	Complete
Embioptera	*Anisembia texana*	None (females and some males), four (males); similar front and back wings	Chewing	Incom-plete
Ephemeroptera	Mayflies	Four (sometimes two); triangular, membranous back wings are smaller than front wings	Vestigial mouthparts	Incom-plete
Grylloblattodea	*Grylloblatta barberi*	None	Chewing	Incom-plete
Hemiptera	Diving beetles, cicadas	None or four; front wings thick at the base or uniformly membranous front wings	Piercing-sucking	Incom-plete
Hymenoptera	Bees, bumblebees, wasps and ants	Four, very rarely none; sparse ribs; front wings larger than back wings	Chewing or chewing-licking	Complete
Blattodea	Termites	Four, rarely none; sparse ribs, similar front and back wings	Chewing	Incom-plete
Lepidoptera	Butterflies and moths	Four scaly wings, rarely none	Sucking	Complete
Mantophasma-todea	*Mantophasma zephyra*	None	Chewing	Incom-plete
Mantodea	Mantids	Four	Chewing	Incom-plete
Mecoptera	Scorpionflies	None or four simi-lar, narrow wings	Chewing	Complete
Megaloptera	Dobsonflies	Four; front and back wings are similar, with abundant ribs	Chewing	Complete
Neuroptera	Lacewings	Four; front and back wings are similar, with abundant ribs	Chewing	Complete
Odonata	Dragonflies	Four similar wings with abundant rubs	Chewing	Incom-plete

Orthoptera	Grasshoppers, locusts	None or four (sometimes small); thick, narrow front wings and folded back wings	Chewing	Incomplete
Phasmatodea	Stick insects	None or four; narrow front wings, folded back wings	Chewing	Incomplete
Phthiraptera	Lice	None	Chewing or piercing-sucking	Incomplete
Plecoptera	Stoneflies	Four; narrow front wings, wide back wings	Chewing	Incomplete
Psocoptera	Book lice	None or four; sparse ribs, front wings larger than back wings	Chewing	Incomplete
Raphidioptera	Snakeflies	Four; front and back wings are similar, with abundant ribs	Chewing	Complete
Siphonaptera	Fleas	None	Sucking	Complete
Strepsiptera	Stylops	None (females), four (males); reduced front wings and triangular back wings	Vestigial mouthparts	Complete
Thysanoptera	Thrips	None or four long, narrow fringed wings	Sucking	Between incomplete and complete
Thysanura	Silverfish	None	Chewing	None
Trichoptera	Caddisflies	Four hairy wings.	Chewing, greatly reduced mouthparts	Complete
Zoraptera	*Zorotypus newi*	None or four long, narrow, fringed front wings; front wings larger than back wings	Chewing	Incomplete

[Source: https://m.espacepourlavie.ca/en/chart-orders-insects]

3.4.5 Insect's Mouthparts

As discussed earlier the mouthparts of insects evolved to serve several different purposes from biting and grinding to sucking and piercing sponsoring and siphoning all sorts of different food products. From the image provided, we can see that the insect mouthparts are made up of the maxilla and labium. Some insects may have a mandible for grinding and biting there are usually highly sclerotized and move at right angles to the body. Many animals formed mouthpieces for haustellating, especially those depending on liquid food, where sucking and siphoning is necessary. Stylets, which are needle-like appendages pierce and penetrate plant and animal tissue.

All Arthropods have a flap -like structure that sits immediately in front of the mouth, essentially acting as the upper lip and forming the roof of the perioral cavity. From the inside, the labrum is membranous and is raised the way from the mandible by two muscles the attached to the inside of the head. The labrum in recent studies has been expressed to be innervated by the tritor cerebellum in the brain and develops as early as from the embryonic period of the insects' life. The inner labrum also contains sensillia (mechanosensilia and chemosensilia) which are essentially sensory neurons that appear like little small hairs or pegs

Insects that chew bite or crush have well-articulated mandibles that can sometimes be used for defense purposes where are the front they have a cutting edge and the molar area has more grinding capability. These jaws can be extremely tough and capable of exerting high amounts of pressure. This is evident in termites and beetles that have shown to have no difficulty tearing even through common metals like copper lead and zinc (Javed & Agurla, 2016).

When masticating, (chewing) behind the mandible is a pair of the maxilla that manipulates and moves the food around. The maxilla sits laterally on each side of the head beneath and behind the mandible. In some unusual cases like in the Dragonfly nymph, the maxilla can extend out and snatch prey back to the head. The labium, the maxilla, and the mandibles are similar in structure as they come from two sides and fuse by the midline of the insect's face along with the median plate. The bottom part of the labial segment including some parts of the Sternum is called the prementum. The preoral cavity is shut from behind by the prementum. The hypopharynx is a membranous lube the projects forwards from the back of the pre -oral cavity and divides the cavity into a dorsal food pouch and a salivarium where the salivary ducts open.

Insect mouthparts that have developed for piercing and sucking have a proboscis and paired mandible and maxillae. The maxillae have stylets that are needle-like in composition and they are enclosed by the labium. When the mosquito is feeding the maxilla with the needle-like stylets penetrates the skin and other parts are then inserted through as it unsheathes. The mosquito then injects saliva that contains anticoagulant through the hypopharynx stopping the blood from clotting and using the labrum it sucks up the already partially digested food (Javed & Agurla, 2016).

Insects that have evolved mouthparts for siphoning have an adaptation of the proboscis where the cibarium muscle in the pharynx is highly developed for pumping action. The proboscis has stylets similar to those evolved for piercing and sucking however they have developed into two long tubes that are held together by a hook but separable. Each tube curves inwards eventually forming a Central tube through which siphoning and suction are possible. This action of suction is done by the contraction and expansion of a small sac located in the head and when at rest the proboscis coils under the head of the insect only extending when it's time to feed. Fascinatingly there have been noted correlations between the thighs and body mass of insects and the length of their respective proboscis.

Some insects like the Honey Bee have developed mouthparts that are evolved responding to food. They have done this by having a well-developed set of ciberial muscles that act in a pump like fashion. Where some features like the mandibles are absent and other structures are reduced the labellum in sponging insects is highly developed having labial palps. The labella are made up of many grooves known as the pseudotracheae that absorb liquids but the first saliva is secreted and used to dissolved and collect the food particles so that they easy for ingestion. The pseudotracheae is known to absorb content within a capillary-like action whereas it can protrude and suck liquid or semi-liquid food. This protractible organ that acts somewhat like a tongue is known as the labial glossae and is hairy so that nectar or honey in the case of honey bees adheres to the hairs and then is retracted so that the liquid is carried further down the food canal (Snodgrass, 2018).

3.4.6 The Insect Thorax

The insect thorax is just as evolutionarily diversified as other body parts but all share similarity in that they are all divided into three segments; the prothorax the mesothorax and metathorax Intersection is / an interesting man to suture and furthermore, each segment has four regions known as;

- The Tergum (or notum) that covers the dorsal end
- Two pleura that sit on each lateral side
- And a sternum on the ventral end

In insect biology, the notum on the prothorax is known as the pronotum, in the mesothorax the mesonotum and in the metathorax, it is called the metanotum.

Similarly, the pleura (pleuron in singular) regions are named according to their respective location on the thorax. There exist the propleura, mesopleura, and metapleura.

The section closest to the head is the prothorax and it hosts the first pair of legs. The dorsal portion of this anterior section is known as the pronotum. This area never bears wings in insects although some fossils showed some projections from this area. In the majority of insects, it is reduced in size but in a few, hypertrophy of this area is possible like in Beatles and some Cockroaches whereby the pronotum extends and forms a shield that covers part of the head in front and the mesothorax behind (Snodgrass, 2018).

The pterothrax (pteron means wing in Latin) is the name given to the middle of the thorax (mesothorax) and the metathorax because this is where the insect's wings are attached. The mesothorax is the middle of three segments wearing the second pair of legs and this is where the wings of all winged insects begin even in cases it may be diminished in size or modified. Some insects like beetles have an elytron which is a sclerotized hardened forewing but what is common among flight- capable insects is the formation of halteres.

Halteres are flight organs that shaped like tiny clubs that give rotational information in the air. They are so important in insect flight that William Dernham who first documented them in 1714 discovered that insects with surgically removed halteres were unable to remain airborne, however, they otherwise behaved normally. It was further determined that halteres are stimulation organs that energized muscles around the Wings and furthermore assist in walking as well as flying in some insects like flies. Their movement also differs in insect species with a variety of wing-haltere oscillations but in general, they offer some compensatory reactions and steering during flight. In some insects, these halteres are also attached to neck muscles which illicit head positions and movement (Steinmann & Zombori, 2012).

A good analogy to offer is with common flies that typically hold their halteres at 90° and to visualize this it is comparable to a man holding their arms out sideways (that would be 180°). If the man then moves their arms backward to a position where the fingertips and spine were right angles to each other it would appear similar to what occurs with the insect, holding their halteres at 90° offset. From this position, the insect can determine motion along 3 axes (pitch roll and yaw).

The mesothorax also has legs in all adult insects and in some species this area may be exaggerated. When this happens is usually around the anterior section of the mesothorax and they would be a sclerotized region that covers the beetles of wing call the tegula. Some insects display a larger mesothorax even than the prothorax, e.g. the Hemiptera order that includes cicada, aphids, leafhoppers, and shield bugs. Here, they form a scutellum which is an enlarged mesonotum. The mesothorax and metathorax have suture lines along their pleura (mesopleura and metapleura sutures) that start along the wing base to the coxa (hip joint) of the leg.

The mesothorax being the middle of the three segments in the thorax hosts the second pair of legs and it is also the segment that the following in all winged insects protrude from even though they may be somehow modified like in Beatles that have sclerotized wink of is known as elytra. They can be reduced even as far as forming strepsiptera or halteres. All insect species have legs protruding from the mesothorax and like the other segment of the thorax, it is also divided into a dorsal -ventral and lateral side known as the mesonotum mesosternum and mesopleuron respectively. In wind insects, there is usually a small sclerite that covers the attachment of the wind to the mesonotum known as a tegula. Some insects like the Hemiptera how the dorsal surface of the thorax covered by an extension of the prothorax (Steinmann & Zombori, 2012).

The metathorax is the last third section or segment of the thorax in an insect and ordinarily hosts The Third pair of legs and just like the preceding 2 sections of the thorax, they are also divided further into dorsal -ventral and lateral sections known as the metanotum metasternum and metapleuron. In wind insects that have more than one pair of wings the metathorax wood host the Hind wings as much as they may be reduced or modified like in Diptera or general flies. In these groups of insects, they are reduced almost entirely to form halteres and in some obviously, they can be completely absent even wear for Rings are still present liking Coleoptera. All adult insects generally have legs on the metathorax and in many, one will find that the metanotum is reduced in comparison to the mesonotum. Some insects would even have the first abdominal segment fused directly to the metathorax and in these situations, they are called propodeum (Shaw, 2014).

3.4.7 General Morphology of Insect Legs

Fundamentally insect legs are organized into the following bits

- Coxa
- Trochanter
- Femur
- Tibia
- Tarsus
- Pretarsus

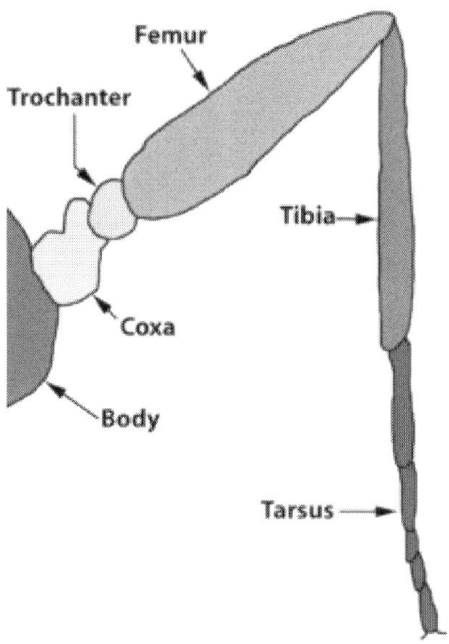

Figure 3.5: A diagram illustrating the setup of insect legs.

[Source: https://en.wikipedia.org/wiki/Arthropod_leg#Insects]

The general similarity of shared ancestry or homologies of legs in insects is difficult to follow and is the cause of many arguments in specialists especially when looking at general anthropoids because many say initially they were up to 11 segments per leg but the most common ancestors may prove to have eight or fewer segments but the ones listed above are easy to follow. You will find that insects have leg segments that are articulated with the next attachment through a hinge joint that only follow one plane, therefore, it means that it needs to have a lot more segments to allow a wider range of movement and like invertebrates that have rotational ball and socket joints that attach the base of the fore and Hind limbs

The Coxa is the most immediate proximal segment and essentially the base of the leg to the various thoracic segments. In comparison, it has the shape of a truncated cone where it may be symmetrical but some have different forms. The back-end of the coxa is articulated usually by the pleuron and is host to two lobes. The posterior lobe is known as the meron and this section is usually the bigger part of the coxa. The coxal corium is the membrane that connects the coxa to the body and the basicosta further strengthens the base of the coxa and gives insight to some of the muscles. These two articulations are the dorsal and ventral points on the coxal hinge and allow movement. On the far the end there are interior decorations towards the Trochanter. There are sutures along the coxa which run along the pleuron to the coxa usually on the outer wall to the Trochanter however it can be absent in many insects.

Another important thing to mention about this section of the leg is the meron which can be found on the posterior lobe and sometimes can be so greatly enlarged that it appears to be a section of the leg on its own and this is evident in some of the more generalized flies like the tipulidae.

The Trochanter is usually a small segment of the insect leg that is flexible along with a horizontal range of the attachments to the coxa but very well fixed almost in movable to the base of the femur. Where this movement is existent along the Trochanter femoral hinge you will find that it is along the vertical plane or somehow oblique vertically with the use of redactor muscles. Both nymphs and adults of the Odonata have two Trochanteric segments that cannot move on each other's plane with the second having redactor muscles that are attached to the femur although usually you will find that most insects have a single Trochanteric segment. Apart from this, there is not much to say about this section.

The femur and insects are the third segment of the leg and are ordinarily the hardest strongest longest part of this limb. It does however vary in size especially when you look at Hind femurs of leaping insects which can be very large and 2 very small in size especially in insects in their larval forms. The size of the femur is directly proportional to the size of the tibia muscle that is inside of it. The typical insect leg has the fourth section as the tibia and it is slender in comparison to the femur but sometimes can be even longer in comparison with a spur that can be 2 or more in number. In insects like the apocrita whisper are usually found along a semi-circular gap in the first segments towards the tarsus and it is lying with small bristles like a cone that the insects use to clean its antenna. Both the femur and the tibia are divided into three sections to represent the front middle and Hind parts of that area respectively being the profemur mesofemur and metafemur for the femur and the protibia mesotibia and metatibia for the tibia.

The tarsus in insects can be divided into 225 sub-segments known as tarsomeres as shown in the diagram but most holometabolous insects especially in the larval form have a more primitive version of this segment. In adult insects, these sub-segments are usually freely movable along with each other and they are articulated by some connecting membranes but don't have any inherent muscles. In some insect, will find that over time as they grow the tarsus although they could be as many as five sub-segments fuse together while and others like the Acrididae they may have the first three tarsomeres United leaving the last two freely movable. The first tarsomere at the base is usually quite enlarged and conspicuous and is known as the basitarsus. Occasionally this segment is also fused to the tibia forming what is known as a tibiotarsus.

Although this is the fundamental morphology of insect legs there are many variations in them for functionality purposes for example many are adapted for running rather than digging while some can be adapted for swimming rather than leaping. The specialized adaptations include four legs for burrowing into the Earth in the Gyrllotalpidae or some for gripping and holding on to prey in Phymatinae there are more commonly known as ambush bugs that sit and wait for prey relying on their amazing camouflage. Some insects like butterflies can have their four legs so greatly reduced that the remaining two pairs are used for walking alone because they are so much adapted to being in the air. Orthoptera is the order of insects that is built up of grasshoppers locusts and crickets and have saltatorial legs built specifically sleeping and in some cases even for defense purposes in kicking. Aquatic insects have legs built specifically for swimming and some have modified some great adaptation so far as to expel water in jets even.

3.4.8 Morphology of Insect Wings

Insect wings are usually found in adult insects and most phylogenetically advanced insects would have two pairs of wings located on the mesothorax and the metathorax. It is interesting to note that insects are the only invertebrates that have evolved in so much that they have developed flight capability and it has been on an Integral part of their success. As much as insect flight is not exactly well understood and it relies on aerodynamic effects that are turbulent you will find that it can be extremely fast and maneuverable mainly due to the way they can change shape. They are however different flight mechanisms depending on the insect order for example butterflies use steady -state error dynamics like the airfoil Theory while others use more complicated dynamics like the Weis Fogh mechanism (Dudley, 2002).

Insect wings are an extension of the insect exoskeleton and have muscles that act along the walls of the thorax that power them when they contract multiple times in a single nerve impulse which ends up allowing the wings to beat a lot faster then what would be thought possible. Internally each Wing has a thin membrane that is built of cells supported by a system of veins. Each of the major veins has nerves and trachea which form of time is the wind develops and dorsal and ventral integument layers form becoming closely lined up side by side forming the general wing membrane (Dudley, 2002). The remaining areas will eventually form future wings where new nerves and trachea will also grow. You will find that wings have two types of hair microtrichia and macrotrichia.

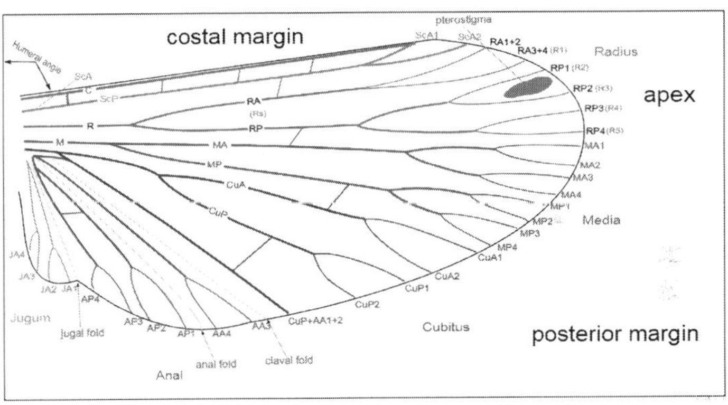

Figure 3.6: Venation of insect wings, based on the Comstock-Needham system.

[Source: https://www.amentsoc.org/insects/glossary/terms/venation]

From the above diagram, one can see how insect wing vernation occurs along the Costa subcostal radius media cubitus and anal veins. The Costa is on the leading edge of the wing and sometimes they may exist a precosta found just right above the costa and you will find it almost never branches like the other veins we mentioned which do. Following that is the subcostal vein and they sometimes may appear to share a trachea. The subcostal vein normally separates into two veins on the anterior and posterior sides. The radial can be thought of as the fourth major vein and is usually the strongest in the wing somewhere around the middle and forks into several different smaller veins (Snodgrass, 2018).

The fifth vein of the wing is the media that folks into two main branches the media anterior and the media posterior which in turn break up into 2 and 4 distal and terminal branches, respectively. The media and the radius share a united stem. The cubitus is the 6th vein that again separates primarily into two branches and this forking in the wing takes place very near the base of the wing with these sectors separating into two more branches but not many more than that even wear it may vary. The postcubitus veins are also known as the anal veins in the Wing system as described by Comstock and Needham and it is very rarely branched if it is mainly into 2.

It is important to note that there are veins that run between each of these major veins connecting them known as cross veins. They run between the Costa and subcosta and connect the radius and media to other adjacent major veins. In some insect orders, the crosswinds are so many that there is a whole pattern that becomes a close link to the network of branching veins and Cross veins that is even difficult to distinguish at times. Rains of insect ordinarily have a convex or concave placement and can be easily seen in some specific insects like mayflies well they fall into a heave pattern.

Wings also contain Fields along what is known as fold lines. Along these Lines wings can actually fold and there also exists flexion Lines which flex during flight and they are very difficult to distinguish from the actual fold Lines. Generally, there are four different Fields in insect wings and they include the remegium which is the biggest area usually separated into two by the median flexion line the Vannal are that is at the posterior anal area the jugal and the auxiliary area which is at the base of the wing (Snodgrass, 2018).

The joint of wings uncomplicated in structure with each wing attached to the Body by a sort of membranous base section containing a number of articular sclerites that are collectively known as ptelaria. There are three distinct plates on the ptelaria and those are the humerus plate at the front side next to the costal vein, the distal plate behind it and the proximal plate next to that. There are also several auxiliary sclerites the first being on the interior hinge plate of the wing base supported by the tergum. The second auxiliary sclerite varies more in form and size than the first auxiliary but still performs the same function as it is obliquely hinged to the outer side of the body of the first auxiliary. This functions as the pivotal sclerite of the wing base and specifically articulates the radial vein.

The 3rd auxiliary is the sclerite where the flexor muscle of the wing is attached and it is variable across all flying insects and is often associated with the base of all of the veins towards the posterior region of the wing and serves as the posterior hinge plate. The contraction of these flexor muscles articulates this sclerite. The 4th auxiliary is not always present in insect wing bases however when it is it is a small plate that sits between the third auxiliary and the posterior notal Wing. There also exist median plates which are also sclerites that are fundamental for flexor muscle operation and the lie away from the second and third auxiliary separated by an oblique line (Steinmann & Zombori, 2012).

Insect wings can have a variety of features like some insect species that have the forewing and the hind wing couples together that improve aerodynamic efficiency. In wasps and bees for example this occurs and there is a role of small hooks on the forward section of the hind wing known as hamuli that hold on to the following and keeps them locked together. When not in flight these wings are usually swept back and sometimes maybe even be folded longitudinally. Some insects also have traverse folding which falls along the flexi that have already been discussed above and they allow sections next to each other of the wing to be folded over or under each other. Although not common this can also be present on the Hind wings like in the Orthoptera where the animal section of the wing usually falls on top of each other like a fan along the lines of the veins. The muscle that allows this to happen or causes the folding mechanism comes from the pleuron through the third auxiliary sclerite.

In relation to flight mechanisms, it is important to recognize that insects have flight muscles that attach directly to the wing where the wind does not beat any faster than the nerves can send impulse commands to the muscles. This only happens in the Odonata and Ephemeroptera orders and other insects that fly use different mechanisms that involve causing the thorax to vibrate for example making the wings beat even faster than what the nerve impulses would be able to do. In terms of aerodynamics, there is the leading edge Vortex or the Weis Fogh mechanisms used with some even capable of Hovering by beating their wings rapidly and having sideways stabilization mechanisms (Steinmann & Zombori, 2012).

3.4.9 The Physiology of Insect Abdomen

The abdomen of insects varies widely with around 11 or 12 segments that are a lot less sclerotized than the head or the thorax. However, you will find that just like the thorax there is a sclerotized tergum sternum and sometimes even a pleurite. Terga (plural of tergum) are separated from the immediate adjacent sternum or pleura with a membrane and usually, the spiracles would be located on the pleural area. For protective purposes, the terga or Sterna together with the terga sometimes may be attached continuously to create both a dorsal and ventral shield and simply a protective tube (Shields, 2017).

Normally the first seven abdominal segments especially in adults' good luck any attachments and are very similar in structure as well. The subsequent areas would have some protruding testicles maybe one or two pairs and evolutionarily they are sort of like coxa and Trochanter lobes in the traditional ancestors' abdominal appendages. Some aquatic insects especially during the larva and nymph life cycle would have gills in this section. The remainder of the abdominal area hosts the reproductive and anal parts. This section of the abdomen is known as the terminally where the anal genital areas around the 8th and 9th segment. The 10th segment is usually free of any appendages however the eleventh segment will have epiprocts or a pair of paraprocts (Shaw, 2014).

The external genitalia that is specifically used for mating or laying of eggs varies widely even in species that appear to be similar in structure. In female insects, the journal internal structures are designed to receive the male copulatory organ and spermatozoa while the meaning of the external structures is used for egg- laying or oviposition. There is usually an egg-laying tube known as an ovipositor in almost all insects but it is absent in some Plecoptera termites and other insects.

3.5 INSECTS INTERNAL STRUCTURES AND MORPHOLOGY

An insect's nervous system is comprised of a brain and a ventral nerve cord like various other invertebrates and this is equivalent to the spinal cord in vertebrates. When discussing the head, we talked about the six fused segments and each of them has a pair of ganglia which are essentially a bunch of nerve cells that's it outside of the brain. You will find that the first three pairs of ganglia are few directly to the brain but the other pairs are attached to the insect's mouth region. Each of the regions on the thorax has at least one ganglion on either side so there is a pair for each segment. This extends all the way to the abdomen but only for the first eight segments maximum. Some insects like Cockroaches have just six ganglia and the abdomen while wasps have three pairs in the abdomen and only two pairs on a thorax.

Not all but many insects have nocireceptors that are specifically a sensory neuron in living animals that respond to damage or potentially damaging stimuli transmitting the sensations of pain. Although these cells exist there is no wide scientific consensus saying that insects can feel pain consciously.

The digestive system of an insect is designed to extract nutrients from the food that they consume and for insects specifically because they are very small they ingest food in the form of macromolecules which is essentially of very large molecule commonly composed of polymerization. The insect digestive system uses a catabolic reaction that breaks these molecules into smaller units that are either oxidized or used in some anabolic reaction for the production of energy. Insects have an alimentary canal that runs lengthwise down the body sending food in One Direction from the mouth to the anus and this is divided into three sections. The foregut, the midgut, and the hindgut (Shields, 2017).

The foregut is built up of chitin and hard fruit in molecules that protect this area from damage when consuming tough food. Here food is chewed in the buccal cavity where digestion starts with saliva and enzymes breaking down the food. Flies and other similar insects have extra-oral digestion that expels digestive enzymes on the food before ingesting it to first break it down and this is a great strategy to allow significant proportions of nutrients to be available from the food source. The midgut is also known as a mesenteron is where most of the digestion happens with microvilli able to diffuse nutrients and use it for the production of energy. The hindgut is where undigested particles are together with uric acid to create fecal matter. This area is very efficient in the absorption of water from the fecal pellets and then the matter is ejected to complete the digestion process.

The respiratory system of insects is done without the use of lungs and they have internal tubes and sucks where gases are actively pumped or diffused transporting oxygen directly to the cells that need it. In insects, you will find that the circulatory system is not used for carrying oxygen and so it is somehow reduced and the insect will not have veins or arteries in the circulatory system. The gas exchange system in insects may vary especially in different groups from continuous to diffuse ventilation and even discontinuous ventilation Cycles that occur normally when the insect is at rest. The discontinuous gas exchange system happens when the insect ticks in oxygen and small amounts of carbon dioxide are released at the same time why the diffusive ventilation is a continuous exchange system of gas through diffusion rather than physically taking in the oxygen (Shields, 2017).

As already discussed the insect circulatory system is not used for the transport of oxygen but rather functions primarily to bathe the insect's internal organs and can be as little as 25% of the insect's total body weight. When we stay with the insect's body organs it could be in hormones nutrients and even the expulsion of waste sometimes for temperature control immunity and the storage of nutrients as well. The circulatory system is also important during the molting process and also can serve as a defense mechanism against predators by churning out malodorous chemicals. Insect blood difference from vertebrae blood because it doesn't have any red blood cells because it doesn't need it but they have something more similar to lymph fluids instead. Because they have a heart organ the operation is similar in that way in that they are contractions usually originating at the insect's posterior end that pump Ford through the dorsal area can out through the hemocoel.

The internal reproductive systems of insects for females can receive and store sperm and also make eggs. They have a pair of ovaries with duct and accessory glands. The ovaries make the eggs and there are accessory glands that help produce and lay the eggs. They also have spermathecae that can store the male sperm for some time as well (Snodgrass, 2018).

Male insects and their reproductive system are built up of testes with the more primitive in simple insects having only one and the more advanced evolutionarily having two mature testes with two separate ducts that supply them. Most insects today however have a pair and inside of them, they are follicles that are wrapped in a membrane. They have an ejaculatory duct that has a cuticle lining the last portion of this area sometimes it is sclerotized to form an aedeagus which is the reproductive organ for the secretion of sperm.

As shown above insects or highly diverse in anatomy and physiology and even as a species. Some insects differ even in the life cycle as you will find that some are ametabolous which mean that they have no metamorphism involved but instead only very between nymphs and adults in size these insects are ordinarily winless and quite primitive. Hemimetabolous insects have a sort of incomplete metamorphosis where the insect Young is somehow similar to adults and have bugs for where the Wings would grow and initially start as naiads. Holometabolous insects undergo a complete metamorphosis which means that they have different physiologies entirely in the immature life in comparison to what they would be in their adult stage. They would also live in entirely different habitats and have very distinct behaviors in comparison as well. They would start first as a larva and even here they would have different phases some turning into in stars and then forming a pupa where does not feed or have any visible activity in this cocoon that they develop they grow wings and the adults finally emergence. A great example is a beautiful butterfly (Snodgrass, 2018).

4

Insect Reproduction and Development

CONTENTS

4.1 INTRODUCTION

The word "insect" is derived from the Latin word *insectum* which means "divided body". Insects belong to the larger Arthropoda family. *Arthropoda* and "jointed legs" are synonyms. Other *Arthropoda* species include crayfish, shrimp, and spiders. One can tell an insect from other arthropods due to the six legs and three distinct body parts – Head, thorax, and abdomen. Other common features on insects include compounded eyes, a pair of antennae, and an external skeleton commonly referred to as an *exoskeleton*. It is hard not to find an insect in every environment. Some of their habitats include soils, caves, the air, water, on the land, and trees. Though small in size, insects are considered the most diverse living organisms since according to studies undertaken in 2017, they have over one million described and found species on the globe. The same study estimated insect population at 5.5m. Of these 1.5m were beetles (Gilbert, 2011).

One of the three body parts needs the other two. They are interdependent and an insect without a thorax is inconceivable. The head is covered by a hard, unsegmented, heavily sclerotized, exoskeletal head capsule known as *epicranium*. It has three main functions gathering food, sensory perception through the antennae and neural integration through the brain. The head may have compounded eyes and antennae where the insects use eyes to spot enemies from far while Antennae help them sense touch, smell, and sound. The brain of insects, though smaller than that of human beings is positioned in the head. The nervous system assembles information from the sensory organs and transports it to the brain where it is assimilated. The shape of an insect talks volumes, one can almost guess what the insect feeds on by just looking it. Those that suck fluids have pointed mouthparts as opposed to the ones that chew food (Heming, 2018).

The thorax comprises three parts *prothorax, mesothorax,* and the *metathorax* in that order from the head. The prothorax bears the first pair of legs and the *pronotum*. While the second pair of legs and the anterior wings are the features of the middle segment the *Mesothorax*. The metathorax segment borders the abdomen and has the last pair of legs and posterior wings. For easy movement, the left of the first pair of legs is located directly across the right leg of that pair. Not all insects have wings. Those with wings are known as *Pterygota* and the wingless Apterygota. Some insects such as wasps and bees have two wings others have four. Some insects crawl all their lives others fly and walk while others swim. Without wings and legs, insects would find it very hard to move about. The abdomen is the largest part of an insect which is softer than the other parts. For most insects, the internal and external reproductive structures are the main features of the abdomen. It is divided into segments whose number varies with insects, for instance, an adult *Collembola* has six, *Zoraptera* ten segments, an orthopteran family *Acrididae* has eleven and an adult Protura-12 segments. It also carries part of the digestive, circulatory, and respiratory systems. All the three parts of the body work for hand in hand one cannot do without the other. An insect minus a head is dead.

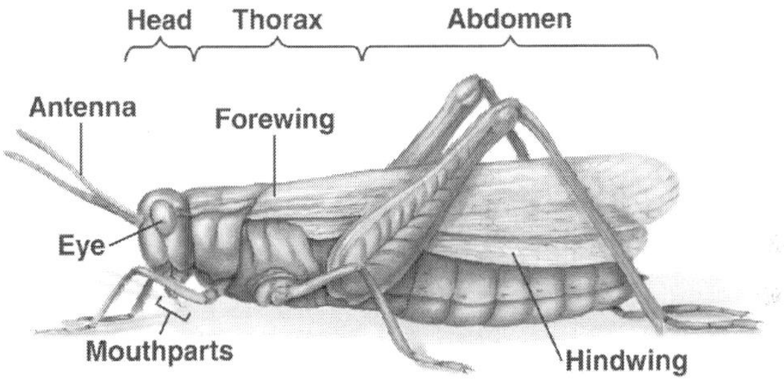

Figure 4.1: A grasshopper showing the three parts and features of an insect.

[Source: https://www.pinterest.com/pin/404690716489195617/]

4.2 REPRODUCTION IN AN INSECT

Reproduction takes place through adult females and males. Let us take a look at the female and male reproductive systems;

4.2.1 Female Reproductive System

The reproductive system of female insects consists of three main features including, a pair of ovaries, accessory glands, *spermathecae* (a sperm pouch), and vessels connecting these parts. The ovaries are made up of numerous egg tubes referred to as *ovarioles*, which vary in size and number by species. While the rate of egg development is influenced by the *ovariole* design, the number of eggs that the insect produces is determined by the *ovarioles*. The reproductive role of a female insect includes the production of eggs, receiving and storing sperms, manipulating sperm from different males, as well as laying eggs. Accessory glands secrets substances that assist in sperm maintenance, transportation and fertilization. They can produce glue and protective substances that coat eggs. Once the female insects receive sperms they are stored in *spermathecae* in the period between mating and egg fertilization.

4.2.2 Male Reproductive System

The male insect either has one testis or two testes that are suspended in the body cavity by *tracheae* and the fat body. The testes has sperm tubes or follicles that are enclosed within a membranous sac. The follicles are linked to the *vas deferens* through *vas efferens* and the two tubular *vasa deferentia* which link to a median ejaculatory duct that leads to the outside. A section of the *vas deferens* is often inflated to form the seminal vesicle, which stores the sperm before they are discharged into the female. The glandular linings in the seminal vesicle produce nutrients that feed the sperm. The ejaculatory duct which stems from an invagination of the epidermal cells during development hence has a cuticular lining. If the terminal portion of the ejaculatory duct sclerotizes, it forms the intromittent organ, the *aedeagus*. Apart from the germ cells or *spermatogonia* which grow from the primordial pole cells during embryogenesis, the rest of the male reproductive system develops from *embryonic mesoderm*.

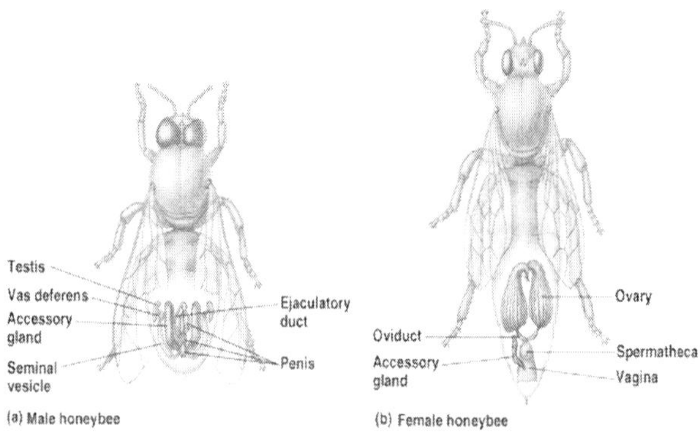

Figure 4.2: Reproductive systems of male and female insects.

[Source: https://slideplayer.com/slide/4404247/]

4.2.3 Types of Reproduction

Sexual reproduction: Among arthropods and insects, sexual reproduction is the most common.

In Sexual reproduction parents (Male and female) produce reproductive cells called gametes which unite to form an offspring. Since Gametes comprise half of the chromosomes found in other cells of the organism, they are referred to as haploid cells. Hap from half. Gametes are products of a cell division known as Meiosis. Fertilization is the process in which two gametes unite. Gametes meet in the process of fertilization out of which a fertilized cell emerges known as a zygote. A zygote has twice the number of chromosomes in a gamete hence referred to as diploid (Marshall Cavendish, 2000).

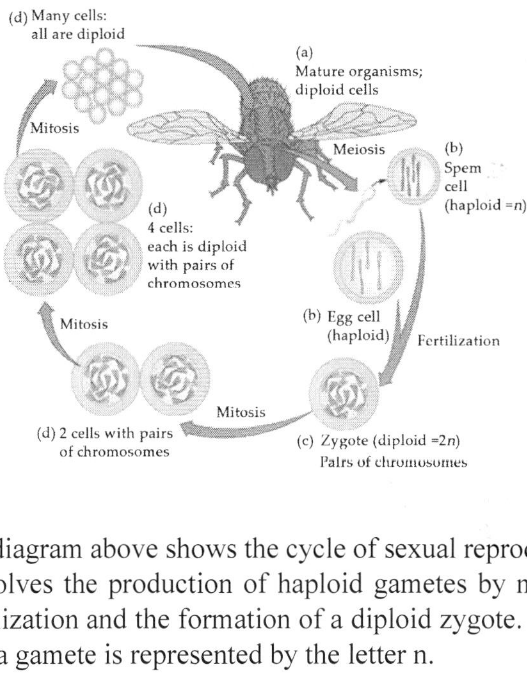

Figure 4.3: The diagram above shows the cycle of sexual reproduction. Sexual reproduction involves the production of haploid gametes by meiosis. This is followed by fertilization and the formation of a diploid zygote. The number of chromosomes in a gamete is represented by the letter n.

[Source: https://schoolbag.info/biology/concepts/56.html]

Courtship: Just like animals and humans, insects reproduce sexually through mating. Before mating, some insects go through courtship. It is hard to think of an insect courting but research shows that they do in a sporadic manner. For instance, male fruit flies (*Drosophila*) are impressive before copulation. On the other hand, blowflies (*Musca*) land on any female and copulate. Males of butterfly and moth descent stay by the pupa and copulate with it once she emerges (Gilbert, 2011).

It is interesting that some females only mate after hearing the voice of a male as in the case of cicadas, crickets, and some grasshoppers. Female grasshoppers that are hard of hearing do not allow mating. Some butterflies *Ageronia* are noisy during courtship. During courtship, male butterflies such as the queen butterfly (*Danaus gilippus*), emits a scent over the female's antennae from "hair pencils" that protrude from his abdomen. No chemical display, no copulation.

The copulation of dragonflies is quite intriguing since the male copulatory organ is found in the thorax as opposed to the abdomen in other insects. A male dragonfly perches on a plant and moves the sperm from the terminal genital opening to the copulatory organ. Then he woos a female and places her on his abdomen and fly away. Once they parch the female receives the sperm from the male. Complete aerial though rare in insects, mayflies, houseflies, ants, wasps, and bees are notorious for this act. Other ways that insects attract mates are through the emission of pheromones, coloration patterns, sounds, and vibrations. The female often stores the sperm in the spermathecae. Once the eggs are ready, fertilization takes place within the female's body, producing an embryo. They hatch and develop into either male or female offspring. Other insects that reproduce sexually include mosquitoes and grasshoppers (Gilbert, 2011).

Figure 4.4: Grasshoppers of the species *Romalea microptera* from the United States, mating.

[Source: http://www.sciguru.org/newsitem/18002/grasshopper-mating-reveals-evolution-decision-making]

Internal fertilization: In this case, fertilization takes place mainly in the female. Offsprings emerge in the following ways through internal fertilization; viviparity, ovoviparity, and Oviparity.

- *Viviparity:* This is where fertilized eggs incubate inside the reproductive ducts of the female hence bear a living young, for example, larvae. The young receive nutrients from the mother's blood through the placenta. The developing offsprings spend a considerable amount of time in the reproduction tract after which the mother releases them to survive on their own with minimal help. Viviparity is common with cockroaches, aphids, scale bugs, and flies (*Muscidae, Calliphoridae,* and *Tachinidae*), in some beetles and thrips (*Thysanoptera*). The majority of the insects' hatch from eggs fertilization and development takes place inside the egg enclosed by a shell (*Chorion*). Some species like *Diploptera* are viviparous and hence incubate inside the mother and are born alive. The following are different types of viviparity in insects;
    - ✓ ***Pseudoplacental* viviparity:** The female insect lays eggs containing little yolk which necessitates her to feed the embryo on a placenta-like tissue. It is common in many aphids and *Dermaptera*, in some *Psocoptera* and *Polyctenidae* (*Hemiptera*).
    - ✓ ***Hemocelous* viviparity:** Embryos grow through obtaining nutrients by osmosis inside the female's *hemolymph* (a liquid similar to blood). It occurs only in *Strepsiptera* and in gall midges.
    - ✓ ***Adenotrophic* viviparity:** Underdeveloped larvae feed on liquids excreted by the mammary glands. The larvae pupate immediately after being laid. This occurs in flies of the families *Glossinidae* (tsetse fly), *Hippoboscidae* (horse or dove flies), *Nycteribidae* and *Streblidae* (bat flies).
- *Oviparity:* Females lay the eggs which hatch after some time as the case of chicken. The embryos feed on the egg yolk. The female lays the eggs in animal or plant tissue, crevices, in soil, and protects them with secretions produces from the accessory glands. To lay the eggs, the female digs up the substrates using an ovipositor (a structure on the last abdominal segment). For

instance, many female grasshoppers lay their eggs in the soil in a frothy matrix. Also, parasitic wasps use long, needle-like ovipositors to drill through the wood to reach host insects. Much as egg production determines oviposition, availability of oviposition sites, and nutrition affects egg production. Females have a coping mechanism food and oviposition sites are available they are prepared to produce eggs. On the other hand, when oviposition sites are rare, oviposition is delayed and this, in turn, slows down egg production. Females focus their energy on survival than producing eggs. However, in some cases, the kind of the oviposition site stimulates egg production, like the cabbage plant encourages egg production in female diamondback moth, *Plutella xylostella*.

Figure 4.5: insects' oviparity.

[Source: https://www.projectnoah.org/spottings/21319105]

- ***Ovoviviparity:*** In ovoviviparity, eggs hatch as they are laid. Females lay fully developed organisms. This is common among the Madagascar hissing cockroach, *Gromphadorhina portentosa*). This occurs in some bony fish (such as the guppy, *Lebistes reticulatus*), some sharks, some lizards, some snakes (such as the garter snake, Thamnophis sirtalis), some vipers, and some invertebrate animals (such as the Madagascar hissing cockroach, *Gromphadorhina portentosa*). Internal fertilization has the advantage of protecting the fertilized egg from dehydration on land. The embryo is isolated within the female, which limits predation on the young. Internal fertilization also enhances the fertilization of eggs by a specific male. Even though fewer offspring are produced through this method, their survival rate is higher than that for external fertilization. Some females protect the eggs from predators by keeping watch.

External fertilization: In this case, fertilization takes place externally without mating. The male deposits its sperm in sealed packets (spermatophores)and leaves them on the ground expecting his female insect partner to pick it up or delivers it to her as a nuptial gift. When the females get this package, they extract this sperm and use them to fertilize their eggs. As it is hard for the females to find the *spermatophores*, some male insects such as springtails encourage the female to pick it up by guiding her to the *spermatophore* with a dance. To guide his partner to the *spermatophore*, the silverfish ties a thread to the *spermatophore* and attach their female partners to it. In a few cases such as that of honey bees, the sperm in the *spermatheca* is always safe for use for the rest of her life (Wigglesworth, 2012).

External fertilization likely takes place when the female lays eggs and the male releases the sperm in the same area and at the same time a process known as spawning. Aquatic species release their sperms into open water which then swim into the eggs in the female's body. Some of the insects that produce spermatophores include *Orthoptera, Psocoptera, Coleoptera, Neuroptera, Mecoptera* and some *Hymenoptera.*

Source 4.6: Orthoptera (female) grabbing the spermatophore laid by a male.

[Source: https://allyouneedisbiology.wordpress.com/2017/11/26/reproduction-insects/]

Parthenogenesis: The term *parthenogenesis* comes from two Greek words, *parthenos* and *genesis* meaning virgin and origin respectively. This is a reproductive strategy where a female's egg produces a new organism without being fertilized. It is also referred to as asexual reproduction. It happens in situations where female insects do not have the capability to mate or in situations where a species multiplication is required. It is also seen as an incomplete form of sexual reproduction since organism parthenogenic organisms develop from gametes. Parthenogenesis is common where the population is low and insect partners are rare as well as when there is a need for mass production of a given species (Locke, 2012).

Parthenogenesis can occur in haploid or a diploid cell. In haploid parthenogenesis, offsprings emerge from haploid eggs to produce haploid adults. This is common in a few species of bees. In diploid parthenogenesis, a haploid cell may fuse with another haploid cell or duplicate its chromosomes to produce diploid zygotes that develop into diploid adults. In another form of diploid parthenogenesis, the diploid egg cell is formed from a parent cell through mitosis (cell duplication). One or more of the daughter cells develop into diploid offsprings. This phenomenon occurs in aphids.

Parthenogenesis occurs in three different forms;

- Arrhenotoky meaning that only males appear from the unfertilized egg cells as observed in wasps, bees, and bumblebees. Two distinct genetic systems are at play- a) haplodiploid-one set of male haploid chromosomes and two sets of female diploid chromosomes. b) diploid, males and females are derived from diploid eggs.
- Thelytoky is an indication that the unfertilized eggs give forth females only which is common among weevils and bagworm moths. This can occur in two genetic systems based on meiosis (automixis) and mitosis (apomixis).
  - ✓ Automixis- This can come about as a result of duplicity of the genome before meiosis creating four chromosomes or fusion of the two nuclei during meiosis.
  - ✓ Apomixis – The absence of meiosis prevents chromosomal rearrangement. The daughter resembles the mother.
- Deuterotoky, both male and female organisms emerge from the unfertilized egg cells characteristic of gall wasps.

Among species, with the haplo-diploid sex-determination system, such as among hymenopterans (ants, bees, and wasps) and thysanopterans (thrips) unfertilized eggs are produced, haploid males. Ordinarily, eggs are laid by the queen but workers who have not mated can lay haploid male eggs on a special occasion. This happens a lot with domesticated honey bees where the queen is the only fertile female in a hive. If she dies without a replacement, then worker bees (Females) to lay eggs. The unfertilized eggs produce only drones (males) since worker bees do not mate. The male mates with a queen only. This can lead to the extinction of a colony. Parthenogenetically, few ants and bees are known to produce diploid female offspring, for example, a honey bee subspecies from South Africa, *Apis mellifera capensis*. Since the worker bees are capable of producing diploid eggs parthenogenetically they can replace the queen in the case of death. Other insect species that can reproduce a new queen include species of the small carpenter bee, (genus Ceratina), parasitic wasps, European formicine ant, Central and South American electric ants, *Wasmannia auropunctata* (Locke, 2012).

Variations: Some variations have been observed including;

Heterogony or cyclic parthenogenesis where generations of organisms formed from unfertilized eggs alternate with those produced from fertilized eggs. This is to some extent attributable to temperature changes. For instance, sexually produced eggs have a greater ability to withstand winter cold.

Pseudogamy (gynogenesis, or sperm-dependent parthenogenesis) This occurs where sperm is required to stimulate the initial development of the egg. Found in the life cycle of a few insects, mites, and salamanders as well as the flatworm *Schmidtea polychroa*.

4.3 SEX DETERMINATION

Sex determination in insects is similar to that of humans. In insects, males and females have a pair of chromosomes that determine an organism's gender. When an embryo acquires a pair of "X" chromosomes it grows into a female referred to as "Homogametic, if it inherits one "X" and one "Y", develops into a male"- known as Heterogametic. The "XX" female is said to be homogametic; the "XY" male is heterogametic. The offspring's gender is determined by the male's contribution. In some cases," like extinction, the "Y" or "X-chromosomes could be missing. If an insect species has no "Y" chromosome at all then females are likely to outnumber the male organisms. For instance, aphids have experienced degeneration of one chromosome during embryogenesis (Locke, 2012).

In Lepidoptera and Trichoptera the sexes are reversed Females are heterogametic (WZ) and males are homogametic (WW). Note that the sex chromosomes have been labeled "W" and "X" to distinguish this system from standard X-Y sex determination. Here, the female's contribution determines the gender of the offspring., these sex chromosomes are designated "W" and "Z" (instead of "X" and "Y").

Top of Form

Bottom of Form

Hymenoptera, many *Thysanoptera*, some scale insects (Hemiptera/Homoptera), and a few weevils (Coleoptera) have a sex determination called Haplo-diploidy. The females are diploid homogametic ("XX") and the male is haploid. Organisms develop asexually. It is interesting to note that the eggs that the female lays can develop into a male. Also, the female can choose to produce a male or a female organism by receiving the sperm through opening the valve of her *spermatheca* to produce a female or blocking the entry of the sperms by closing the valve of her *spermatheca* to produce a male. Some insect communities limit the production of males to reduce competition for limited available resources and increase the production potential of the population. A good example here is bees, wasps, and ants which have more females (queens and workers) and males are produced for reproduction as need be (Reznik, Voinovich & Vaghina, 2009).

4.4 INSECT GROWTH

The life cycle of most insects starts with eggs which vary in color, shape and size laid singly or in groups but are protected by a coating that protects the embryo from and prevents water loss. The eggs could be a barrel-shaped, round, oval, disk or suspended on long stalks.

Once the insect hatches the offspring enters the immature stage. It grows to a point where the exoskeleton cannot stretch further. A new cuticle grows under the old one and then the old exoskeleton splits and the insect moves out. This process is referred to as moulting and the phase between different molts is an instar. The new exoskeleton hardens shortly after exposure to air and growth in the insect resumes. Insects change their form to a varying degree with each moult. Molting takes place many times over varying periods until the final stage is reached. It stops once the insect matures into an adult (Heming, 2018).

4.4.1 Metamorphosis

For the majority of the insects to mature, they undergo *metamorphosis*. This process is controlled by a juvenile hormone secreted by glands in the insect's head. This hormone is released during moulting. If it is introduced in the wrong amount and the wrong time during the insect life cycle, then normal growth can be disrupted. Agriculturalist/farmers apply this principle to control pests. Metamorphosis could be gradual where change is subtle or complete metamorphosis where every new stage is distinctly different from the previous one. Insects are classified into three groups by entomologists based on the metamorphosis they undergo: *Ametabolous, Hemimetabolous*, and *Holometabolous*. Insects adapt to climatic conditions. For instance, when the weather is unfavorable, they go into dormancy a state known as diapause (Heming, 2018).

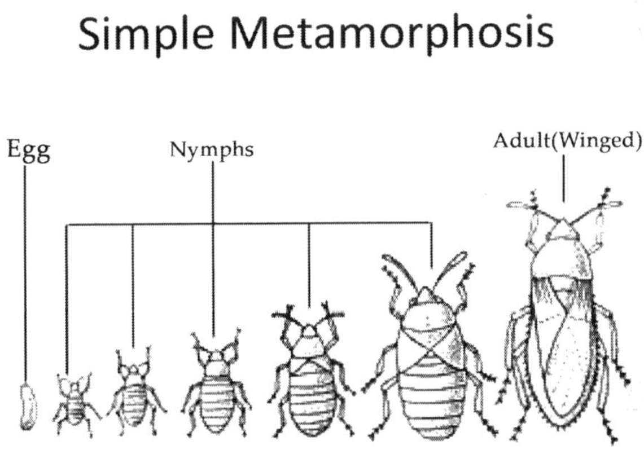

Figure 4.7: Simple metamorphosis.

[Source: https://slideplayer.com/slide/12651072/]

- *Ametabolous: Little or No Metamorphosis: Ametabolous* is a Greek word meaning "having no metamorphosis". The most ancient insects, such as apterygotan (wingless) insects such as Collembola, springtails, silverfish, and firebrats, undergo little or no true metamorphosis during their life cycles. Once these insects hatch, the immature insects are similar in appearance to the adult. The young insect (nymph) resembles a small adult. The reproductive structures are undeveloped in the nymph. They moult progressively until they reach maturity. Through moulting, the insect develops in size and sexual organs.

- *Metamorphosis or Paurometabolous Development:* This is a less primitive form of metamorphosis common in cockroaches, grasshoppers, mantis, and white ants. The young (nymph) look like the adult in body structure, behavior and habitat through the wings and the reproductive organs are undeveloped and small in size. In the early stage, the wings grow as wing pads in *mesothorax* and the *metathorax* thoracic segments and slowly develop in size by mitosis in each moult. The genitalia also grows with each moult. All other body parts develop as the young moults. The nymph leads an independent life and attains adulthood through several moults. This type of metamorphosis is referred to as gradual metamorphosis since the nymph undergoes consistent growth in each moult until it becomes an adult.
- *Incomplete Metamorphosis or Hemimetabolous Development:* Entomologists generally refer to insects that undergo this metamorphosis as, "*Hemimetabolous*". Hemi means "part" and regards this type of transformation as incomplete metamorphosis. The life cycle shows three stages: Egg, Nymph and Adult. For the *hemimetabolous* insects, growth usually occurs at this stage. Wings bud. Nymphs and adults resemble in appearance and behavior aside for the size. They share the same food source and habitat. A few hemimetabolous insects include grasshoppers, mantids, cockroaches, termites, damselflies dragonflies, and all true bugs.

4.4.2 Insects that Breed in the Water

Most insects that breed in the water or are aquatic undergo incomplete metamorphism. The nymphs, in this case, are called naiads. They respire by gill. Naiads like terrestrial nymphs, molt, and in stature until they become adults when they are released from the water. The wings and genitalia are developed when they attain adulthood. For example, when a mayfly leaves the water it rests on a tree and undergoes one final molt to become an adult.

Naiads have three pairs of legs, a head with a compound eye, antennae, and a small abdomen posterior tracheal gill. The features of an adult include; a smaller head, a larger abdomen, and breathes using spiracles. Adulthood life span varies with insects; mayfly lasts a day.

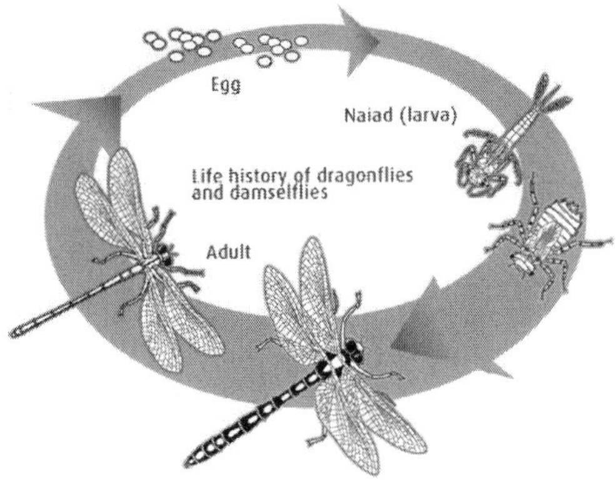

Figure 4.8: Hemimetabolous metamorphosis in dragonfly.

[Source: https://www.pinterest.ca/pin/222294931577495954/]

4.4.3 Complete Metamorphosis or Holometabolous Development

Entomologists have named to insects that undergo complete metamorphosis "*holometabolous*," from "holo," meaning "total." These are insects that go through a four-phase life cycle, from egg, larvae, pupa, and adult. Most insects undergo complete metamorphosis throughout a lifetime. Each stage of the life cycle— egg, larva, pupa, and adult. Each stage is marked distinctly. The insects in each stage look different for the other. An egg does resemble larva; neither does pupa nor adult. They use different sources of food from the adult.

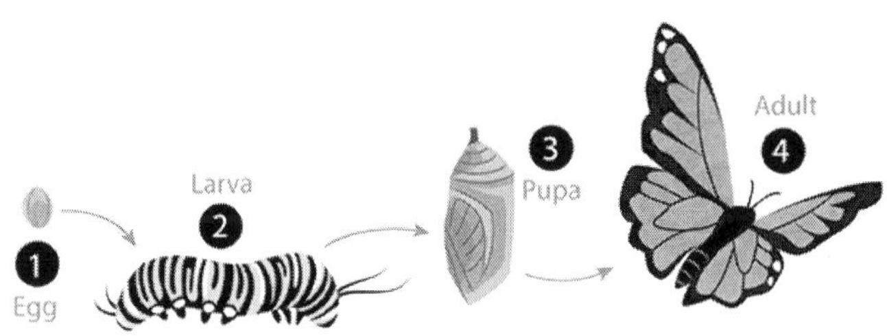

Figure 4.9: A complete-metamorphosis.

[Source: https://askabiologist.asu.edu/complete-metamorphosis]

Egg stage: The female lays the eggs in animal or plant tissue, crevices, in soil, and protects them with secretions produces from the accessory glands. The mother lays the eggs with the feeding of the larvae in mind. For instance, butterflies lay their eggs at the back of cabbage leaf which acts as feed for larvae. The outer coating of the egg (Chorion) is tough enough to protect the embryo. The Chorion comprises *chorionic plastron* which holds a thin sheet of air close to the surface of the egg and ensures a supply of oxygen into the egg. The egg bears the genetic information required for the growth into adulthood including the blueprints for imaginal discs. Imaginal discs become the anatomical parts of the adult form. For instance, out of the egg emerges compound eyes, *carapecs*, mandibles and exoskeletons.

Larvae Stage: Larvae stage if the eating and growing stage. In an insect experiencing complete metamorphosis, it is the immature stage between the egg and the pupa.

The following are the types of larvae; oligopod, polypod, and apodous.

Oligopod Larvae: This larva has no abdominal legs but the thoracic ones are intact. It is divided into the following sub-types;

- *Campodeiform*: They derive their name from their semblance to dipluran genus *Campodea*. It has a well sclerotized, depressed dorsoventrally and elongated body. Prognathous head and long thoracic legs. It has a pair of abdominal cerci or caudal processes. They are active predators. Examples include the grub of ladybird and antlion.
- *Scarabaeiform*: It has a well-formed head, Sub-cylindrical, stout and C-shaped body. Caudal processes are absent and short thoracic legs. It is slow and digs into soil and wood, for example, the grub of rhinoceros beetle.
- *Polypod* or *Eruciform:* It has an elongated body and a sclerotized head bearing a pair of powerful mandibles with which it feeds on vegetation. Either side of the head constitutes two groups of single lensed eyes (Stemmata). It has three pairs of segmented thoracic legs which have claws used to cling onto leaves. It also has five pairs of prolegs and the bottom of these legs (planta) which has rows/circlet of short hooked spines useful in to clinging to the smooth surface of the vegetation and walking. Three to six abdominal segments, for instance, caterpillar the larvae for moths and butterflies. The larvae have;

- ✓ Hairy Caterpillar – The body is hairy – Hair is densely, sparsely distributed or arranged in tuffs, eg, Red hairy caterpillar.
- ✓ Sluggish caterpillar: The larvae is fleshy, stout, thick, short and poisonous spines (scoli) distributed all over the body. The head is small and retractile. Bears tiny thoracic legs. The abdomen is not distinct and the abdominal legs are absent. For example, platform larva.
- ✓ Semilooper: It has three or four pairs of feeble prolegs in either the third and fourth abdominal segments. A good example is the castor semi looper.
- ✓ Looper: Also known as measuring worm, earth measurer or an inchworm. It only has two pairs of prolegs in the sixth and tenth segments. For example, Daincha.

- Apodous: They are larvae without appendages for mobility. Are based on the sclerotization of head capsule and degree of development. There are three sub-types;
    - ✓ *Eucepalous*: This larva has a well-developed head capsule, operating mandibles, maxillae, stemmata and antennae. Mandibles act transversely. For example, Wriggler (larva of mosquito) and grub of red palm weevil.
    - ✓ *Hemiccephalous*: The head is reduced. Mandibles are verticle. Examples include Larvae of horsefly and robber fly.
    - ✓ *Acephalous*: It does not have ahead. A pair of curved hooks are found in the mouthparts. They are also known as vermiform larvae. Example Housefly larvae maggot.

Prepupa Stage: This stage occurs between at the end of the larvae stage and the beginning of pupa stage. Here feeding ceases and sometimes a cocoon is produced. The prepupa resembles the larva but it is often shrunken and less pigmented.

Pupa Stage: The pupa is immobile and stays in a cocoon to protect it from predators. It is often referred to as a phase of internal reorganization / the resting phase. A pupa neither eats nor grows in size. Minimal external changes occur but active changes occurring internally, hidden from view. Most of the larval tissues such as muscles, epidermis, and the nervous system are re-specified. It is at this stage that the wings are fully grown. This kind of development is referred to as *endopterygota*. Once the reorganization is complete and the environmental conditions are suitable, then the adult emerges.

In the arthropod family, only insects go through complete metamorphosis. The following insect groups go through a complete metamorphosis; the *Endopterygota*, *Endopterygota* includes 11 Orders, the largest being *Diptera* (flies), *Lepidoptera* (butterflies and moths), and *Hymenoptera* (bees, wasps, and ants), and *Coleoptera* (beetles).

There are three types of pupae: exarate, *obtect* or *coarctate*.

- Exarate pupa: It has legs, appendages and an abdomen that is capable of movement. It is found in almost all insects aside for Lepidoptera. Also commonly referred to as "free pupa".
- Obtect pupa: It is characterized by wings, and appendages stuck to the body by a molting fluid. *Obtect pupae* are compact, with the legs and other appendages enclosed. It bears a slender stalk at the top which attaches it to twigs. This type of pupa is observed among moths and butterflies. The pupa of butterflies is known as a chrysalis.
- *Coarctate* pupa: A hard barrel-shaped chitinous case, called puparium encloses the house fly pupa. Externally the puparium is segmented and the spiracles are projected outwardly. *Coarctate pupae* develop inside the larval skin. The pupa of mosquitoes contains broad anterior cephalothorax and is comma-shaped.

A pair of small respiratory trumpets are found on the dorsal side of the thorax. Numerous hairs guard the openings. The abdomen has nine segments and bears a pair of paddles on the eighth segment by which the pupa swims. Once the pupa is fully grown it emerges by bursting or dissolving the pupa case. After a period of pupal existence, the young insect emerges out by breaking or dissolving the pupal case. Coarctate pupa is observed among dipterans (Wigglesworth, 2012). *Coarctate pupae* develop inside the larval skin.

4.4.4 Hypermetamorphosis or Hypermetabolous Development

This metamorphosis is seen in the insects that have a few distinct features. For instance, beetles.

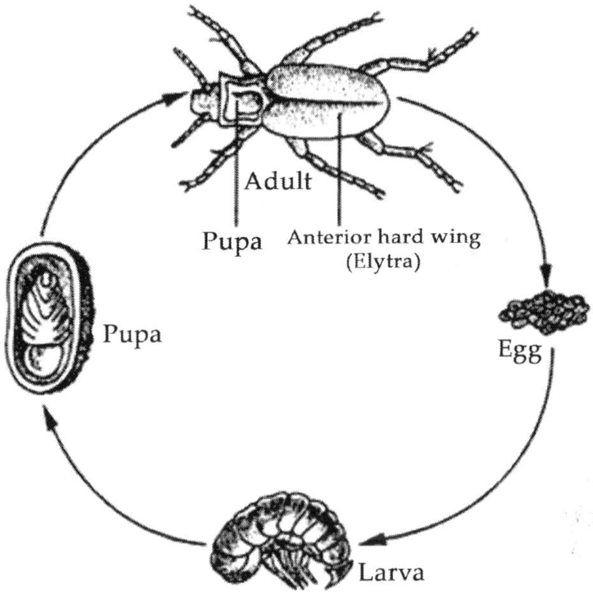

Figure 4.10: Illustration of hypermetabolous development in Blister Beetle.

[Source: http://www.biologydiscussion.com/invertebrate-zoology/insects/metamorphosis-kinds-events-and-role-of-hormones/33522]

At the end of the pupal stage, hormones are released which softens the shell wall and allows the adult insect to emerge. The emergence of the adult is referred to as *Eclosion*. This being a new environment, the adult insect remains immobile acclimatizing to the weather conditions. The winged ones cannot fly since the wings are still wet and folded. They lie there at the mercy of their maker. At this juncture, the adult is at risk of predators. Once the wings dry up the winged ones fly, some crawl and others walking (Locke, 2012).

At this stage, the insect is fully grown male or female with six legs and three body parts, head thorax, and abdomen. They cannot grow any further. The adult's main function is to reproduce -mate and lay eggs. The female insect moves from place to place in search of a suitable place to lay her eggs. Some insect males move around in search of a partner.

Insect Generations: The length of a generation varies with weather conditions and in different insects. For instance, aphids, have many generations each year and keep reproducing as long as the weather is favorable. Some have one year-generations while others like the beetles and moths need two to three years to complete one generation.

Complete *metamorphosis* helps insects divide roles between various stages of the life cycle. The stages are independent of each other hence changes take place in a stage without affecting the previous or next stage, for instance, the feeding of the larvae does not affect the adult stage. Also what happens in the larva stage is not dependent on the pupal stage (Locke, 2012).

4.4.5 Events of Metamorphosis

Research has revealed during egg formation, gametes in an egg are grouped into two– A group to be used up during the larvae stage and the other group to support pupal and adult life. As the larva cells grow, cells develop in size but never divides. Another set of cells known as imaginal buds and discs, remain inactive in the larva. Once the larva is fully grown, the imaginal buds take full charge. At the pupal stage, imaginal cells grow by sub-division. At the cellular level, the pupa is inactive (Gilbert, 2011).

During metamorphosis, most of the larval organs in the pupa except the central nervous system and developing reproductive organs are broken down by enzymes and the process of disintegration of the larval organs is called histolysis and these larval disintegrated cells die and are used up by the imaginal cells.

In certain insects, within larva, pupal cells become fluid in consistency and imaginal cells continue to form the adult structures. The imaginal buds are the groups of formative cells that remain inactive in the larva stage forming the rudiments of future organs by mitosis.

These formative cells set aside in the pupa and reach functional organs by differentiation in the imago (adult). The process of formation of tissues and organs from the imaginal buds are referred to as histogenesis. The wings, mouthparts, internal organs, muscles and legs develop from the imaginal buds (Gilbert, 2011).

Insect reproduction takes place in a wide variety of rather complex forms. Some insects oftentimes change the type of reproduction during their life cycle in response to climate change. For instance, during winter in the temperate regions, some insects are capable of activity, others migrate to warmer areas, while others have evolved mechanisms of diapause that allow eggs or pupae to survive these conditions. The rest produce the same way throughout their lifetime.

4.5 FACTORS THAT AFFECT INSECT REPRODUCTION AND GROWTH

Change in climatic conditions affects insect's feeding behavior, fertility, distribution, intensity either positively or negatively. Researchers placed controlled hatching of Asian lady beetle (*Harmonia axyridis*) under different temperatures, 41, 39, 37, and 25 Degrees Celsius. After some time, they observed that those placed under 25 Degrees Celsius had hatched sine that is the optimal hatching warmth. None of the eggs that were placed under 40 Degrees Celsius hatched while very few of those placed under 39 and 37 Degrees Celsius hatched. (Khaliq et al, 2014).

When it is too cold or too hot, then the survival of the insects overshadows reproduction. Extreme temperatures are not conducive for the growth of the eggs which thrive best in warm moist conditions. This is dependent on the type of insect and the habitat. For instance, in the tropics termites and ants hibernate in holes during the rainy season but are very active in dry weather conditions. When an area experiences prolonged drought, then insects lack food, some migrate, others hibernate in holes while others die leading to a reduction of their population in a given area. During winter in the temperate regions, some insects are capable of activity, others migrate to warmer areas, while others have evolved mechanisms of diapause that allow eggs or pupae to survive these conditions. The rest produce the same way throughout their lifetime.

The static or inactive state is referred to as dormancy. Dormancy occurs during different stages of the life cycle. For example, some damselflies survive as larvae frozen in ice during winter, many butterflies and moths as pupae and grasshoppers as eggs; and some mosquitoes survive as adults. This explains why many butterflies are noticeable in the period that follows the rainy season – Autumn. In the event of extreme weather conditions, insects such as locusts, grasshoppers result in migration(Khaliq et al, 2014).

- High temperatures influence the insect life cycle stage, growth, and internal metabolism. For example, the hatching period for *Helicoverpa armigera* egg was observed for about eight days at 28 degrees Celsius, when eggs were placed under 25 degrees Celsius, the hatching period went up by three days. Insects do not bear very high and very low temperatures, for example, studies showed that Aphids can reproduce 5% more with an increase of temperature by two Degrees Celsius.

- Vegetation: The habitat of herbivore insects is important in population fluctuation. For instance, bushes are a good breeding ground for mosquitoes. Insect feeding and multiplication was higher on grazed pasture.A
- Crowding: Crowding of a given insect population species can lead to competition for resources. For instance, if a queen would boost egg-laying this would lead to an increase in population density.
- Human activity: As the population grows communities tend to encroach onto the available idle land. In most cases the forested areas as in the case of Kenya – Encroachment in the Mau Water tower. To settle in such areas, they fell trees to make room for their homes. In so doing, they interfere with the life cycle of living organisms in the forests for example butterflies, killing some in the process reducing their population and cause others to relocate.
- Pollution: As countries industrialize, pollution is becoming a big problem since poisonous gases are emitted that are hazardous to life. It affects the ecosystems in many ways such as global warming causing climate change. Plants absorb some of the gases and some dry. This affects the life cycle of insects in that their food is reduced and this leads to slow reproduction. Herbivorous insects can be checked by pollutants affecting the ecosystem. An increase in organic and inorganic pollutants affects the terrestrial ecosystem. This has led to a reduction of pupal weight, an increase in growth larval mortality and accumulation of metals such as nickel, cadmium, zinc, and copper as they ingest plants laced with these metals. In cases where refuse is drained into rivers then the life cycle of the aquatic insects is interfered with.

5

Insect Senses and Communication

CONTENTS

5.1 INTRODUCTION

Senses are certain parts of our bodies that allow us to perceive various things. For instance, eyes are used in perceiving sight, mouth perceives taste, ears perceive sound and the skin perceives touch. Though these senses are in human beings they are also present in animals more so insects and is a useful resource. It helps them protect themselves from danger and also helps them to look for food. This also places emphasis on communication. Communication refers to the sending and receiving of information from the sender to the receiver. Insects communicate with each other as they interact with each other in their activities. Some insects like human beings have a system that governs them. Take the case of bees who have a queen that governs the hive. According to the type of bee, they have different roles to play. To ensure that all these activities are done effectively there is a need for communication among the bees. The study of insect's body parts, as well as their sensory organs, is known as insect physiology (Heming, 2018).

Insects have different types of senses that will allow them to smell, hear, touch, and taste things in their environment similar to those of human beings. This is implying that insects have eyes, mouths, ears, and skin. There are experimental results that indicate that insects have sensory capabilities. These capabilities may be different from humans either quantitatively or qualitatively. These organs work as sensory organs or receptors. They then perceive the various elements in the environment and become transducers. They take the role of converting chemical energy, mechanical energy and light energy from the environment to convert them into electrical energy of the nerve impulses within the sensory neurons. The generated signals are transmitted to the brain (the ventral nerve cord) where they lead to the stimulation of the most suitable behavioral response. According to the type of stimuli produced the brain develops the appropriate behavioral response.

Some of the behavioral responses may include looking for resources such as food or mates, reacting to the changes in the environment or avoiding danger. The sensory receptors are different but are all from the embryonic ectoderm. They are also integral parts of the insect's exoskeleton. Depending on the role they play they can be divided into three groups. The groups include chemoreceptors, photoreceptors, and mechanoreceptors (Heming, 2018).

Mechanoreceptors: These receptors perceive changes in the mechanical aspects of the environment. Some of the examples include the tactile receptors, proprioceptive receptors and the sound receptors. Some of the roles that these receptors play include the detection of movements within the environment. They also perceive vibrations and other mechanical disturbances. It is most useful in protecting the insect from harm more so insects that tend to be prey for other organisms.

Chemoreceptors: These receptors perceive changes in the chemical composition within the environment. Some of the examples include taste buds, palps, sensilla and the antennal. Their role is to detect the presence of chemical substances present in the substrates that are in the taste or in the air in terms of smell. This assists the insects to migrate to environments that will allow their survival as there are environmental chemical components that will limit the survival of certain insects.

Photoreceptors: These are receptors in charge of perceiving changes in light concentration. Some examples include the compound eyes as well as the Ocelli. Their functions include detecting the presence as well as the quality of the electromagnetic radiation commonly referred to as incident light. To get a better understanding of how insects are able to perceive these senses, it is required to have comprehensive knowledge of the nervous system of an insect. The nervous system allows the transmission of the obtained impulses or signals from the sensory organs of the insect and thereby transmitting the impulses to the brain where they are interpreted allowing the most suitable response to be carried out.

5.2 INSECT NERVOUS SYSTEM

It has been said that insects usually have a very complex nervous system. Their system makes use of a variety of external sensory information as well as internal physiological information. In the nervous system, the basic component is the nerve cell also referred to as neuron. It is composed of a dendrite having two projections. These two projections have a role of receiving the stimuli. It also contains an axon that transmits the received stimuli to another neuron such as the case of a muscle. There are chemicals referred to as neurotransmitters that are given off at the synapse. Some of the neurotransmitters include dopamine and acetylcholine (McGregor, 2005).

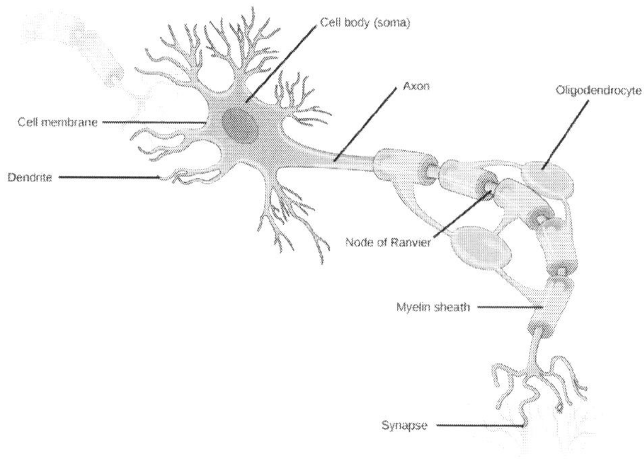

Figure 5.1: A dendrite used in the transmission of impulses.

[Source: https://courses.lumenlearning.com/wm-biology2/chapter/neurons/]

Insects like human beings also have a central nervous system. This is the most important part that is useful for the sensing and production of an impulse. The central nervous system works alongside the endocrine system.

5.3 AN INSECT'S CENTRAL NERVOUS SYSTEM

It is composed of the brain, sub- esophageal ganglion, and a ventral nerve cord. The sub-esophageal ganglion is connected to the brain by two nerves that extend outwards. An insect's brain is composed of three lobes that are the protocerebrum, deutocerebrum and the tritocerebrum. The protocerebrum innervates the compound eye as well as the ocelli. The deutocerebrum innervates the antennae while the tritocerebrum innervates the labrum and the foregut. The ventral nerve cord is an extension of the sub-esophageal ganglion posteriorly. Insects have a layer of connective tissue known as the neurolemma that covers the brain. There are also the ganglia which cover the important peripheral nerves and the ventral nerve cords.

In the head of an insect, there is a head capsule that is composed of six pairs of ganglia. It is structured in a way that has the first three pairs of the ganglia fused into the brain while the remaining three parts are fused to the sub-esophageal ganglion. The thoracic segments on each side have one ganglion. They are further connected to a pair and hence one pair per segment. For most insects they have fewer numbers of the ganglion attributed to fusion or reduction. In the case of cockroaches, they have six ganglia in their abdomen. The ganglia in the nervous system play the role of coordinating centers that coordinate the impulses in given regions within the insect's body (McGregor, 2005).

To assist insects process the impulses received at the sensory organs, they make use of the peripheral nervous system that is composed of the motor neuron axons. These branch out from the ganglia to the muscles. Sensory neurons and also the sympathetic nervous system. The sensory neurons are those from the circular sensory organs. These organs receive different kinds of stimuli such as the thermal, mechanical, visual or chemical stimuli from the environment. The sympathetic nervous system is composed of the ganglia and nerves (McGregor, 2005).

5.3.1 Sensory Organs

Here we discuss in depth the knowledge on the sensory organs of the insects.

Taste: In line with the chemical senses we have the chemoreceptors. They are for taste and smell. Reception of such stimuli affects various aspects of insect life such as their mating, feeding, habitat selection, and their parasite-host relationships. The mouth is mostly related to the sense of taste. Most insects make use of their mouthparts to perceive this sense. Some other insects make use of other body parts such as their antennae to perceive taste. The reason is that for such insects, their taste organs are located at the antennae. Insects with this attribute include the wasps, bees, and ants. For other insects such as moths and butterflies, their taste organs are located in the tarsi.

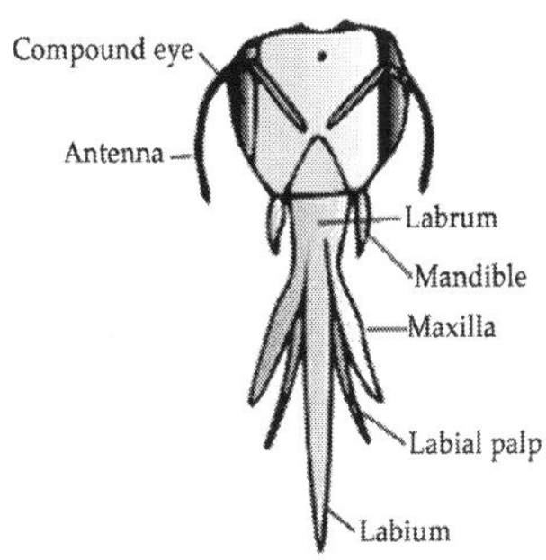

Figure 5.2: The mouthparts of an insect.

[Source: https://www.amentsoc.org/insects/fact-files/mouthparts.html]

Smell: Insects make use of the olfactory sensilla to allow them to smell. It is usually located in the antennae. For some insects, chemoreceptor sensitivity is related to the smell in some substances. This shows the difference between some insects as some can perceive a given smell though they are at very low concentrations and may be far away from the source of the smell.

Motion: Motion tends to revolve around the mechanical senses. With such senses, the insects can obtain information such as the movement of trees in the environment or even the location of their enemies. Such senses affect the general movement of the insects, reproduction, and flight from enemies as well as feeding. Insects can perceive these senses through obtaining stimuli such as pressure, vibration and touch. Insects have hairs on their cuticles that are responsible for the insect's sensitivity to variables such as sound and touch.

Hearing: The hearing organs in an insect are located in different parts of their bodies. They are also known as tympanal organs. For some insects, they are located in the abdomen, wings, legs or antennae. They can respond to a variety of frequencies that range between 100 Hz to 240 Hz. This factor is dependent on the type of insect species. Insects have joints that are composed of tactile setae. This registers movement. They also have hair beds that are useful in the determination of the proprioception which is also the information concerning the position of the limbs. They are located in the cuticles found at the joints of legs and segments. If there is pressure applied on the strain gauges or body wall of the insect, it will be detected by the campaniform sensilla as well as an internal stretch receptor sense muscle distension.

Vision: Insects make use of their compound eyes and the ocelli to provide for vision. The compound eye is composed of the ommatidia. This refers to the light receptive units. The number of compound eyes of an insect varies according to the type of species. For instance, ants may have two or one compound eyes while dragonflies are said to have ten thousand compound eyes. There is a striking relationship concerning the ommatidia. This is that the more the number of the ommatidia the higher the visual acuity. The light receptive units are made up clear lens system as well as light -sensitive retina cells. The retina cell is composed of rods and cones one of which allows the perception of color (Kaner, 2002).

Figure 5.3: An insect's compound eyes.

[Source: https://www.quora.com/Would-humans-have-a-greater-advantage-if-they-had-compound-eyes]

As insects obtain vision, the image obtained is composed of a mosaic of specks resulting from differing light intensities obtained from the different ommatidia. There are case scenarios whereby the visual acuity is forfeited for light sensitivity. This is in cases of dusks or night time. This is because at such time the ocelli are unable to form focused images. The images tend to have a higher sensitivity to the difference in light intensities. Insects are also able to perceive color though their view is better in the blue end of the spectrum compared to the red. Still, in some orders of some insects the sensitivity range may be composed of the ultraviolet.

Temperature sensors: Most insects have humidity and temperature sensors. In most cases, the insects that are smaller in size tend to cool quickly in comparison to larger animals or insects. Insects are generally considered to be cold -blooded that is they are exothermic as their body temperature will fluctuate depending on the temperature of the environment. An exception is made on the flying insects as they can raise their body temperature. This is made possible through the action of flight that goes above the environmental temperature. For instance, in the case of moths and bumblebees, their bodies are insulated by scales and hairs. For such insects, during flight they can raise their muscle (flight muscle) to between 20 to 30 degrees Celsius above the environmental temperature.

For other flying insects, they are required to maintain their flight muscle at a certain temperature to allow them to attain the required amount of energy or power to allow them to fly. For larger insects, this is facilitated through activities such as shivering or vibration of the wing muscles. This helps them to increase the temperature of their flight muscles allowing them to fly (McGregor, 2005).

Recently it has been discovered that there is the presence of cells known as nociceptors that facilitate the detection as well as the transmission of sensations of pain in insects. This theory tries to point out that there is a possibility that insects do feel pain. The sensory organs play a crucial role in the communication that occurs between insects as it forms part of the social insect communication.

The insect senses play a major role in insect communication. The senses all work together to aid in the communication between insects which is important. Depending on the type of sense the insect communication will be dependent on the type of sensor used by the insect. They vary depending on the areas of sensitivity. These areas are specialized for perception. For instance, insects such as bees can perceive ultraviolet wavelengths. They are said to have the ability to detect polarized light. In the case of moths, they make use of the antenna more so the males allowing it to detect pheromones of the females' moths over distances of several kilometers (Kaner, 2002).

In the case of the yellow paper wasp, they are known to make wagging movements which are considered as a form of communication within its community. The wagging can generate frequencies that can be considered as communication. In the instance of communication, the wagging movements may signify the arrival of new material into the nest. In the case of aggression present between workers may be used in the simulation of others to increase the foraging expeditions. There is a modification of the insects when they utilize a certain area of their body. This can be demonstrated by a trade-off between visual acuity and chemical or tactile acuity such that when an insect has well- developed eyes, they tend to have simple antennae.

There are different ways in which insects can perceive sound. Insects have the ability to produce sound though they vary in the patterns in which they perceive the sound. The insects have different ways in which they hear sounds. They usually hear a narrow range of frequencies that are in relation to the number of sound frequencies they produce. This can be demonstrated by mosquitoes and grasshoppers. Mosquitos have a frequency of 2 kHz while grasshoppers can hear the sound to a frequency of 50 kHz. Based on sound, the insect can detect the presence of prey or hosts based on the properties of sound they produce. A good example is the nocturnal moths that can perceive ultrasonic emission made by bats. This enables them to avoid predation (McGregor, 2005).

For some insects that are blood feeders, they make use of their special sensory structures to allow the detection of infrared emission so that they can make homes on their hosts. The solitary wasps are known to display a rudimentary sense of number to allow them prey on single species. The mother wasp lays eggs in individual cells and givers the individual eggs with a number of live caterpillars which will be fed on by the eggs once they hatch. One of the ways in which insects communicate with others on the basis of the type of sense they utilize. One of the ways that they communicate is through the use of visual communication.

5.4 LIGHT PRODUCTION AND VISION

One of the senses present in insects is sight. It gives them information through the use of vision. Their brain gets information through vision and the brain interprets the information obtained. Most of the insects such as fireflies among other species of insects have the ability to control the light generation through the production of flashes. They use this tool differently for some, they use it to attract mates while others use it in attracting prey. Some of them have glowing abilities that may be used in the attracting of a small flying insect into strands of silk. Some of the fireflies mimic the flashing of the female species to allow them to attract the male species that are later on captured and devoured. The colors emitted from light vary from dull blue to rare reds (Hedwig, 2013).

For some insects such as those of cave crickets can perceive light and dark. Most of the species have acute vision capabilities that allow them to detect even minute movements. The insects have eyes that include simple eyes or even compound eyes which may vary in size. With the different sizes, the insects are modified to be able to detect light in the infrared, visible and ultraviolet wavelengths. The eyes of insects may be similar to those of human beings. The difference may be in the Rhodes where they can perceive certain colors in the environment.

Part of the visual communication in insects involves insects communicating through the use of visual signs. The color patterns and markings present on insects such as butterfly wings as well as moths. This is important as it is used by insects in recognition as they can be identified by the colors present on the bodies of the insects that are useful in their identification. For some insects, they make use of bright colors and distinct features that may be utilized in scaring predators away. For some insects, they utilize certain dance movements so that they can attract mates or facilitate the communication with their nestmates. These movements play a significant role in communication so long as they are done in broad daylight where they are visible to insects (McGregor, 2005).

This poses a challenge and to counter fit it is that some insects have the ability to generate their own light allowing the signs to be visible at night. On regards to signs, there are things known passive signs which include eyespots or color patters which are utilized as a form of advertisements. Most of the appropriate actions done are active signals that are given off by insects. They include body movements and light flashes which are used in appropriate times. They are said to contain higher information content in the signal frequency, duration or the period in which the signals are conveyed as they present different meanings. Different species pose unique flash patterns as well as the response time.

There are species of insects that produce J-shape flashes when they are in the rising flight movement. The female may respond with single flashes in a two -second interval. In some cases, the role may be reversed. For instance, in tropical seasons the females fly while the males will perch in the vegetation. Different from human beings, insects have the ability to see ultraviolet light. This allows the species to communicate through the use of various wavelengths in this spectrum. For instance, the female cabbage butterflies make use of ultraviolet reflecting scales present on the dorsal wing surface. When in-flight the down strokes of the wings result in the creation of a flash of UV which is recognized by the males as a flight signature of a potential mate (McGregor, 2005).

In the case of vision, both beetles and flies can make their own light. The fireflies are a unique breed of beetles that have the ability to make their own light. In this case, the female flies sit on the ground while the male flies fly above them. Depending on the species of the male firefly flashes their own signal of light emitted from their bodies. The females then search for the signal from the male fireflies that are of the same species. The female fly then sends feedback to the male. In some instances, the insects make use of colors in visual communication. For some of them, they make use of colors in their bodies in the mating process as they show themselves to the females. Some of them make use of ultraviolet light that is not visible to humans. They make use of ultraviolet patterns in the communication process. These are some of the ways in which nature allows them to prepare themselves for activities such as reproduction, in short during the mating process.

In the case of alfalfa butterflies, the males are the ones that have UV-reflective surfaces. In the case of their courtship, the males put a flickering display before the females. In the case of missing scale, there is a reduction of the wing reflectivity which is an indication of aging which reduces the chances of a male getting a mate. Though this method of communication is useful, it has its advantages and disadvantages (Brumm, 2013).

5.4.1 Advantages of Visual Communication

There are various advantages of using visual communication. One of the advantages is that this method of communication is effective when used over long distances. This facilitates long- distance communications. It is also beneficial in that the signals that are transmitted are well understood and the information is well transmitted. The other advantage is that this mode of communication can be utilized as the insects are in motion. This allows them to do several tasks altogether. This method of communication is convenient as it is fast. It makes use of light which travels at a faster pace therefore it may travel at the speed of light.

This method is effective as it can be used in multiple directions meaning that it is independent of factors such as wind. This reduces the impairment in this mode of communication. In the case where passive signals are used, there is minimal utilization of energy. This is because passive signals have minimal energy requirements.

5.4.2 Disadvantages of Visual Communication

This method of communication requires a clear line of sight. If there is an obstacle, then the image will not reach the insects' eye. The other disadvantage is that the visual signals may be interrupted by predators. This method is only effective in the daylight as light is required for the insects to view the signals. It is in the exception of fireflies who have the ability to provide their own light. The other disadvantage is that the active signals may be relatively expensive to produce.

5.5 TACTILE COMMUNICATION

For insects, this is an important form of communication. Most of the insects have poor vision and sound perception. To overcome this challenge, they use physical contact to allow them to communicate with each other. This is why the phrase "keep in touch" is most relevant to insects. For instance, in the case of blister beetles, the courtship process begins with a series of antennal taps done by the male on the sides of the female's body. In return, it responds by lifting her wing covers,therefore, allowing the male to climb on her back. In a bid to complete the quest, the male is required to continue tapping. The tapping is done by alternating it from side to side at a given frequency perfect to facilitate the simulation of her genitalia and therefore initiating the mating process (Kaner, 2002).

In tactile communication, it is important for the antennal tapping to be done. It is mostly used by both termites and ants. Though it is not well understood how the information is sent, this type of communication involves a nest-mate recognition and may result in the exchange of food through the use of trophallaxis. In some cases, there is instantaneous feedback which allows the sender to monitor the receiver's response. It may allow the altering of signals if required. The tandem running in ants and termites involves the antenna tapping on the hind legs. It is commonly referred to as a follow the leader behavior. The tapping leads to the informing of the leader that she has not lost her discipline.

In the case where the tapping stops the leader is inclined to instinctively turn around and look in an ever-widening circle. This is done until she re-establishes contact with the follower. In the tactile communication system, it involves the use of dance language. In the case of honey bees, the dance is performed in total darkness. It is done on a vertical surface of the honeycomb. The dance in this case is referred to as a round dance. This dance involves sending of signals to the nest-mate. It notifies them the presence of a nectar source within close proximity of the hive. It may involve a series of circular runs having more or less frequent changes in the direction taken. In the case where there are greater frequencies than the more the changes in direction and therefore the better the quality of the source of the nectar. In the case of longer distances, the dance used is the waggle dance. It involves the use of an eight pattern having a series of abdominal waggles on a straight run after the half-circle turn (Swanson, 2013).

In order to have a better understanding of the tactile communication, it is important to know more about dance languages. A dance is denoted by the duration of the straight run as well as the frequency of the waggles. The direction is usually indicated by the angle of the straight run. It also corresponds to the horizontal angle present between the sun and the direction of the food source. The tactile cues are usually generated by the ripples on the surface of the water, therefore, allowing whirligig beetles to monitor the location of the dozens of other nearby whirligigs frequently. Due to this tactile communication system, the whirligigs can swim quickly in circles. This is done without them bumping into each other. This is done while allowing them to detect the presence of possible prey or predators.

For certain families such as the treehoppers, they are able to produce vibrations in the tissues of their host plant which can be felt by another treehopper present in the same plant. The signal will travel through the plant. It is best demonstrated by the way banging water pipes in an apartment leads to the creation of noise through the whole building. In some of the species, the signals work like an alarm system. For instance, in the case of nymphs to elicit protective maternal behaviors. In the case of small insects, there are substrate vibrations that allow for an effective communication system. This is more so useful in small insects who cannot generate an acoustic signal that is loud enough to be heard more than a few inches away (Swanson, 2013).

5.5.1 Advantages of Tactile Communication

This mode of communication allows for instantaneous feedback. This means that the relying of information is done quite fast allowing a better response when it comes to the identification of nectar or in the case of possible preys. This method is also used in a localized area and is therefore used among the same species hence effective communication. Since this type of communication involves contact between the two insects, it, therefore, allows individualized receiving of information. This type of communication can also be effectively used in dark areas such as caves among others. It is therefore effective as it is independent of light as it mostly utilizes the sense of touch.

5.5.2 Disadvantages of Tactile Communication

This method of communication is not useful for long distances. It may require the insects to be close to each other within a given range for effective communication. The other disadvantage is that it requires that the insects should be in direct contact with each other for the signals to be effectively relayed. It is therefore dependent on the closeness of the organisms during the communication process. This method requires that the message should be repeated to the recipients. This method of communication involves the individual transmission of signals therefore the messages have to be repeated to the insects that are to receive the information. The other disadvantage involves the vibration signals being intercepted by predators this means that the information may not be able to reach the involved insects. Apart from the use of antennae in communication, there is also the use of the mouth in communication as a form of touch communication. This method is more so useful when insects reside in dark areas allowing them to recognize other insects. When using their mouths, the insects also make use of chemicals in ensuring communication.

5.6 CHEMICAL COMMUNICATION

There are various ways in which insects make use of chemicals in the transmission of messages. One of the ways in which they communicated involves the use of chemical messages such as pheromones and allomones. This type of message involves the insects giving off certain smells also known as insect perfumes as a form of a message. In the case of pheromones, these are chemicals that are used by insects as they communicate with insects of the same species. For instance, in the sex pheromones, they are used in the attracting of insects of the same species for mating. In the case of aggregation pheromones, these chemicals are used in the sending of messages to insects that are within one species. It involves the use of alarm pheromones which are signals that are put out by insects when threatened or disturbed. This form of signals makes ants run out of their tunnels and all over the grounds when the nest is disturbed (Hedwig, 2013).

The trail phenomes are utilized by ants and caterpillars among others. The signals take the resemblance of the maps which assist insects in finding food. It involves the first insect finding the food. It then leaves a trail of scent for the colony. The colony then follows the trail so that they can find more food. Another term used in chemical communication is allomones. This is one way in which insects send warning signals to insects of other species. In some cases, the allomones can be very smelly and in some cases may cause blindness to the animal it is directed at.

It has been noted that most of the insects tend to rely on chemical signals compared to other forms of communication. The signals used include semiochemicals and info chemicals which serve as a form of language that facilitates the mediation of interactions that are present between organisms. Insects usually vary. Some insects are highly sensitive to low concentrations of the chemicals. In other cases, the few molecules may be sufficient to elicit a response (Hedwig, 2013).

The semiochemicals can be divided in to two groups. This category is done on the basis of the group that sends a message and the group receives the message. The semiochemicals are divided into pheromones which are chemicals that carry information from one individual to another member of the same species they include the sex attracts, trail marking compounds, and alarm substances among other intraspecific messages. There is also the allelochemical which are signals that are known to travel from one animal to another member of insects which is of a different species. Such chemicals includc rcpcllants, compounds utilized in the location of suitable host plants and a variety of other substances that are used in the regulation of interspecific behaviors (Kaner, 2002).

The allelochemical can also be divided into three groups based on who benefits from the message. They include the allomones which in this case benefit the sender. They include the repellent or defensive compounds which facilitate the detection of predators. There are also the kairomones that benefit the receiver. They include odor used by parasites in the location of hosts. Finally, there are the synomones that benefit both the sender and the receiver. They include plant volatiles that attract insect pollinators.

Insects can utilize their sense of taste or smell in the detection of semiochemicals. The specialized receptors may be located in different areas of the body more so in the feet, palps, antennae, and ovipositor. The sense of smell is utilized in the remote chemoreception thereby allowing the detection of some semiochemicals used in low molecular weight that are volatile enough to facilitate it being airborne. The sense of taste is used in contact chemoreception allowing the detection of molecules that adhere to a substrate or to the outside of the insect's body (Wyatt, 2003).

5.6.1 Advantages of Chemical Communication

This method of communication is not limited to environmental barriers. This means that the message can be effectively relayed to the intended recipient. This method is also effective when used over long distances and also around corners. As the environmental factors cannot affect it, it makes this method to be eligible for use over long distances. The fact that it is dependent on the senses also allows this method to be used over long distances. The other advantage is that this method can be utilized both in the day and also in the night. This method of communication is longer lasting compared to visual and auditory signals. This makes it more efficient for use. This method is advantageous in the aspect of metabolism in that this method is metabolically inexpensive as only little amounts of the chemicals are required for this type of communication to be furthered.

5.6.2 Disadvantage of Chemical Communication

This method of communication involves the use of chemicals and it may not quite present the information that is to be presented. The scent may only indicate that is there is a message to be transmitted but not the actual message that is to be transmitted. This method of communication is also not effective in an upwind direction as it will lead to the disruption of the signals being transmitted.

5.7 SOUND COMMUNICATION IN INSECTS

Another way in which insects communicate among themselves is through the use of sound. For sound, the sense of the body that is utilized is the ears. For some insects, the organ of sense that is mostly utilized in communication is the use of ears. This means that they have hearing sensors are the most important part of the communication process. In the use of sound in the communication process, it involves the production of sound and the sound being perceived by other insects through the use of ears. The number of ears present on the body of an insect varies depending on the frequency of the sound that has been produced (Brumm, 2013).

Insects can make sounds that allow them to attract their species or send out relevant signals to other species. In the case of insects, they utilize supersonic sounds. The supersonic sounds used in this case are relatively high such that the human ear may not be able to perceive it. This is because the body of the insects is modified in a manner that allows then to detect the slightest movement in the environment. Their ears are also modified in a manner that they can pick up the slightest vibrations and they can hear noises from distant places. This allows them to quickly identify the presence of potential predators and they are therefore able to protect themselves. For some species of insects, they rely on the use of echo in the identification of food. In this case, their bodies are fitted with sensors so that they can identify the echoes that have been sent back.

There are various ways in which insects produce sound. For some of them, they produce sound by rubbing parts of their body together. This process is known as stridulation. In the case of grasshoppers, they can make this noise by rubbing their legs against their wings. Such kind of noises made by insects cannot be perceived by human beings. In the case of scientific research, some instruments are used in amplifying these sounds. There are other ways in which the insects can make sounds. For instance, bees buzz. This type of sound is loud enough and can be heard by human beings. The buzzing is produced as an action of their bodies. The buzzing can be used to kccp prey or to notify the colony on various things (Brumm, 2013).

This type of communication is also known as acoustic communication. When the sound is produced by the animal, it is distributed and disturbs the nearby molecules resulting in a compression of waves that travel in different directions but away from the source. The sound waves are then perceived as by the eardrums thereby causing a mechanical disturbance which results in the causing of a mechanical disturbance which is detected by the sensory neurons present in the inner ear. For the case of insects, the use of sound in communication has been known to be well effective. There are three variables used in sound communication which include frequency, amplitude, and periodicity.

The combination of all these three variables can be used in the creation of an extremely wide and complex range of signals which are utilized differently in insects. The acoustic signals can be fast started, stopped or even modified in the aim of sending a time-sensitive message. There are some insects that can detect sounds that are above the human frequency. This implies that they can produce and detect sounds above the human hearing frequency. The study of insects is known as entomology. When researching with regards to sound, it may involve the use of audio transducers which is a device that facilitates the conversion of inaudible high- frequency sound into low audible frequency (McGregor, 2005).

For most insects, they make use of the tympanic membrane which is present in the abdomen in the case of grasshoppers and moths in the case of other insects it is the tibiae which is present at the front legs. In the case of mosquitoes, they have antennal hairs which allow them to resonate with given sound frequencies. For some insects, they utilize sounds that pass through solid objects. This is insects such as ants, bees, treehoppers, and termites among others. The sounds from the objects can be sensed by insectsvia substrate vibrations with mechanoreceptors in their legs. Like in other forms of communication there are advantages and disadvantages of this type of communication.

5.7.1 Advantages of Acoustic Communication

This type of communication is not affected by environmental barriers. This means that it is conducive for use in the given environmental settings of different insects. This method of communication is used and effective for use in long distances and around corners. For this form of communication, it is highly variable and allows fast change. This is all-important in allowing the transmission of high information content.

5.7.2 Disadvantage of Acoustic Communication

Some of the disadvantages include the fact that this method of communication may not reveal the location of the sender to the potential predator. This method is also less effective in noisy environments for instance in the seashore. This method is also known to be metabolically expensive as it will require the production of sound that is to be heard from fairly larger distances. This method is an attenuation this means that the intensity falls with the distance from the source which is also known as the cube-root function.

5.8 VIBRATIONAL COMMUNICATION

This mode of communication is wide- spread among insects in their interactions. For most of the insect species that make use of sound in communication, then this mode is the most common. It is used alongside other mechanical signaling mechanisms. The vibrational signals utilized usually vary which some having low frequencies, others pure tones or in some cases they make use of a combination of acoustic sounds. For insects, they make use of the vibration produced by plants among other things present in the environment for their benefit. It allows them to locate food as well as predators among others. The insects also modify these vibrations to allow them to be used during the communication process (Brumm, 2013).

The vibrational communication involves some type of mechanical disturbance that propagates through a medium. Some of the mediums include airborne and sound. There are also substrate vibrations are used in insect families. In some insects, it is known as vibrational signaling. For humans, these vibrations may not be heard but some instruments may be used in amplifying the vibrations. The insects use a variety of vibrational signals in every way. For some of them, they make the vibrations through the abdomens so that they can make a variety of sounds. To explain further, it occurs such that the insect utilizes the muscles in its thorax and abdomen to shake the abdomen which in turn vibrates. The produced vibrations are transmitted to the stem and through the insect legs. The vibrations, later on, travel out in both directions along the stems and the vibrations are picked up by another insect which may be a meter away or still present on the same plant.

For them to be able to pick up the vibrations, insects such as treehoppers have sensitive legs which allow them to be able to detect the slightest vibrations. These vibrations can also be used by insects in warning them. They are modified in a way that the other insects can detect the sense of danger. These vibrations are of a given frequency different from other vibrations making them unique. Once these vibrations are sent the insects may be identified of the presence of predators. These vibrations are also utilized by other insects in communication during mating so that the males can identify the presence of the females and vice versa. For the winged insects, their wings can be utilized in the making of vibrations. The wings may be flapped together resulting in production in vibrations (McGregor, 2005).

For some of them, they make these vibrations by touching the surfaces of water which leads to the production of vibrations, and in this case water is the medium for transmission for the vibrations. All the forms of communication used by insects work together in the insects. Some insects have all the sense of communication but some of them lack some of the senses used in communication. They,therefore, utilize the available senses.

To make the communication process possible, there are various parts of the insects' body used in the communication process. In the case of visual communication, there is the use of eyes (McGregor, 2005).

5.9 INSECT EYES AND COMMUNICATION

Insects like other human beings have eyes. The difference is that the eyes of insects are more complex compared to those of human beings. Some of the insects have compound eyes while others have simple eyes. For the insects, some of them have eyes that are made up of several tiny light detectors that are packed close to each other. The presence of several eyes allows the insects to spot very fast movements. It also allows them to have a wide -angle of view. This is useful in helping them to avoid prompt attacks from hungry predators (Swanson, 2013).

For insects that have compound eyes, they are the type of eyes that contain several mini eyes joined together. The individual tiny eyes have lenses that allow them to collect the light. They also have sensors that will allow them to pass information to the brain of the insects. Some of the insects having this kind of eyes are the flies. The compound eyes are huge and take up most of the space on the head of the insect. This means that they stick out to the sides allowing the insects to view things from either side as well as behind it and in front of it. The large compound eyes allow the insects to view the slightest changes in the environment.

The insect has bumps on its head. These bumps are also called ocelli which are simple eyes that allow the insect to pick up movement as well as the changes in the environment. The name compound eyes come from the repeating units also known as ommatidia each of which functions as individual visual receptors. The ommatidia usually entails certain parts. They include a lens which is usually present at the front surface and is known to make up a single facet. There is also a transparent crystalline cone. There are also the light-sensitive visual cells that arranged in a radial pattern and finally, there are the pigment cells that are used in separating the ommatidia from the neighboring parts (McGregor, 2005).

The pigment cells are used to ensure that the light entering the ommatidium is parallel to its axis reaches the visual cells and also triggers nerve impulses so that the information is transmitted to the brain. The ommatidium is pointed at single areas in space and leading to the contribution to the information about one small area in the given field of view. In a compound eye, there is the presence of several ommatidia and the facets are spread over the surface of the hemisphere. In some insects, their eyes have relatively fewer ommatidium such as in the case of grasshoppers while others have relatively more ommatidium compared to others. This is similar to the case of honeybees and dragonflies.

However, even though there may be poor resolving ability though they have several mini eyes. With the use of compound eyes, there is a phenomenon called the flicker effect. The compound eye is known to be exemplary when viewing objects more so those that are in motion. Usually, as the object moves across the eye of the insect, the ommatidium is usually turned on and off thereby resulting in the flicker effect. This effect allows the insect to view objects better when in motion compared to when the objects are stationary. Take the case of the honey bees, they tend to visit the windblown flowers compared to the ones that are readily still. When looking into the eye, we look into resolution and sensitivity. This is the aspect where we look into the ability of insects to be able to apt and be active in dim light. This allows them to concentrate on the screening of pigments of their ommatidia in the lower ends of the pigment cells (Swanson, 2013).

The shift facilitates light entering into a single ommatidium thereby stimulating them. The presence of several ommatidia that are responding to a single area in the visual field allows the image produced to become coarser. In this casc of praying mantis has the ability to do more than distinguishing dark and light in the evening. There is the shift that occurs in the pigment which makes it more sensitive to light than it is in the day time. This is more than the ommatidia can detect while in a given area of light. There is also color vision. This aspect deals with the ability of insects to distinguish between colors. This type of viewing makes use of two pigments each of which can absorb light in different wavelengths.

In the case of the honey bee, there are four visual cells in the individual ommatidium allowing them to respond better to the yellow-green light. They both respond optimum to the blue light. The other two respond best to ultraviolet light. This system makes it conducive for the honey bees to be able to differentiate between colors in the exception of the color red. This has been showing been shown by behavioral studies. Some insects make use of ultraviolet vision. They make use of the ultraviolet-transmitting lens. This has been well demonstrated by the insect- pollinated flowers where the colors of the flowers appear different from the honey bees while compared to how the human beings view it. The ultraviolet vision may not be limited to other animals (Swanson, 2013).

When discussing compound eyes, there are two types of compound eyes. They include apposition eyes which are further divided into two groups. A basic composition eye has a lens that allows the focusing of light obtained from one direction on the random while the light from other directions is usually absorbed by the dark wall present in the ommatidium. The other form of the apposition eye is found strepsiptera that have lenses allowing the formation of images. The produced images are later on combined in the brain. The second category is the superstition eyes which is divided into three types which include refracting, parabolic, and reflecting superposition eye.

The refracting eye has a gap that is present between the lens and the rhabdom and has no sidewall. The individual lenses interact with the light which is at an angle on the other side. The resultant image is usually half radius of the eye. This is where the tips of the rhabdoms are present. This type of vision is utilized by nocturnal insects. The parabolic superposition is usually located on the inside of the individual facet focus light which is from a reflector to a sensor array. For some species of insects such as the honey bee have specialized zones of ommatidia that have been organized into a fovea area that allows for good vision. The acute zone is usually flattened and the facets made larger (Wyatt, 2003).

The flattening allows for more ommatidia to receive light from a given area resulting in a higher resolution. Though most of the insects have compound eyes, some of them have simple eyes. There also some that have a single lens compound eye.

5.9.1 Simple Eye

This is also a different type of eye that is present in insects. It is most likely that these eyes are found in developing insects as most of the adult insects have compound eyes. This eye is also known as the ocelli. These eyes are different from the compound eyes in that they have a single structure instead of multiple ones found in the compound eyes. Some of the insects that have simple eyes include the nymphs and dragonflies. Insects having these type of eyes usually have small swellings. In some instances, insects may have two to three simple eyes. These types of eyes are not utilized in distinguishing between visual images. They are instead utilized in the detection of the changes in light.

There is also the stemmata which are also known as the lateral ocelli. These eyes are usually present in the larvae stage as the insects develop. The insects while in these stages rely on the stemmata which are usually made of a single ommatidium. The term lateral ocelli come about because they are visual organs that are usually located on the sides on the larva's head thereby allowing the recognition of light changes, color, form and distance (McGregor, 2005).

For some insects, they have photoreceptors that are present in some organs which allow the detection of light more so with ultraviolet light. For some of the insects, they have the photoreceptors in the genitals which is very useful in the mating process. This type of vision is utilized by butterflies. In the case where the photoreceptors are damaged (one pair) they will still be allowed to mate. In the case where the photoreceptors are damaged in the female then the communication process may be damaged. All these types of eyes are very useful in the communication process whereby vision is utilized. All these elements are very crucial for proper viewing of elements in the environment facilitating communication in insects more so dealing with the dances (McGregor, 2005).

5.10 INSECT TASTE IN COMMUNICATION

Another part of insect senses includes the sense of taste which is found in the mouth. There are insects that utilize their legs also in testing. In some cases, one might say that insects have taste buds which is common to insects. Insects such as butterflies make use of their legs in testing. Their legs are made of taste buds that allow them to locate the appropriate areas for laying of the eggs. The taste buds also allow the insects to communicate. Another part of the insect's body used in sensing is the mouth. The mouthparts are usually rolled-up in a straw- like manner. Insects have different mouth parts that may be utilized in communication. There is a primitive arrangement of the mouthparts that may be used in making sounds (Wyatt, 2003).

They may also be utilized in the collection of food, processing, biting and also making sounds that will be used in communication. Some of the mouthparts include the labrum which is a cover that may be loosely referred to as the upper lip. There are also the mandibles which are hard powerful jaws that are used in the cutting of food among other things. There is also the maxillae which can be referred to as pincers. These mouth parts are less powerful compared to the mandibles. These mouthparts are used in the steady manipulation of food. They consist of five segmented palps termed sensory and often concerned with taste. There is also the labium which is the lower cover.

It is also known as the lower lip. It is a representation of the fused pair of the ancestral second maxillae. It is divided into three segments palp and is also part of the sensory. Lastly, there is the hypopharynx. This is a tongue- like structure present on the floor of the mouth. It also contains the salivary glands which discharge saliva through it. The mouthparts are modified differently in different animals. The insect mouthparts demonstrate some kind of parallel evolution. For some of the evolutions have been independently attained. The modifications on the mouthparts vary depending on the species of insects. They include the honey bee where the mandibles are relatively smaller and are adapted for molding wax. The labium is curving downward and inwards resulting in a tube used by them in sucking nectar (Wyatt, 2003).

In butterflies and moths in the modification of their mouths, it has involved the mandibles disappearing. The maxillae in their case are elongated and are channeled along their surfaces. They are held together by hooks and spines resulting in the formation of a sucking tube. In the diving beetle larvae, their mouths have been modified to do piercing and sucking. The mandibles are each curved over and it forms almost closed groove along the inner surface. The insects may be modified so that they can allow the production of sounds that are used by insects in the communication process. With the production of sound then there is a need for ears which will allow the insects to perceive the sound (Brumm, 2013).

5.11 INSECT EARS AND COMMUNICATION

When discussing the insect ear parts one has to look into the ear structure so that they can get a better understanding of how insects can perceive sound. It starts with the sound that is produced causing a disturbance in the air particles. The disturbance is transmitted through the air reaches the eardrum. The membrane is also known as the tympanic membrane. The disturbance causes a vibration by the eardrum which is then transmitted in the ear structure. The vibrations produced by the eardrum translate the pressure waves of sound to three different bones in the body known as the hammer, anvil, and the stirrup bones. They then shake against a snail shell-shaped organ commonly known as the cochlea. The curves of this cochlea are covered with hairs. There is a wiggling bone that is outside it jiggle inside the organ and leading to the triggering of the cells present in these organs (Swanson, 2013).

When these cells are triggered, they send signals to the brain through the auditory nerve. The cells are usually hair cells respond to the high frequencies which are closest to the origin of the wave that is being propagated. The cells are also responsive to low frequencies which are located deep in the cochlea. In insects, the tympanic membrane is connected to a thin cuticle plate, and the incoming sound waves hit the cuticle plate like a seesaw which is later on lopsided. This happens while the fulcrum is closer to one end. This structure allows the translation of the larger movements of the air pressure waves to smaller and more powerful motions present in the cuticle plate.

The plate later on results in the creation of ripples in the fluid -filled chamber akin to an unfurled cochlea. Here in this chamber sensory cells are arranged in a similar manner of the keyboard which is from high to low-frequency sensitivity. The cuticle plays the role done by the hammer, anvil and stirrup present in the mammalian ears. These ears have also undergone evolution thought they years to allow the insects to listen more to the environmental changes in matters sound. The modification in the ears allows the insects to be more equipped in the environment so that they can listen to the slightest changes made in the environment. This allows them to properly detect the presence of predators or prey. In matters of communication, the ears allow them to listen to the sounds and vibrations that are made by their fellow insects (Wyatt, 2003).

Inscets have been well adapted so that they can be able to communicate with each other allowing them to signal their colonies and mates concerning the presence of food and predators among others.

6

Insect Social Behavior

CONTENTS

6.1 INTRODUCTION

Many different insect species can be found in the category of social insects. These insects normally live in colonies and have three main characteristics: the normally have overlapping in their generations, they divide labor, and they have group integration. The two main social insects are ants and termites, however, various bees and wasp species fall in this category.

We normally differentiate social insects through behavior, function, and structure, where the main ones are those that have the purpose of reproduction, such as the queen, and then there are the sterile that are mostly soldiers and workers. When it comes to the function of the reproductive class, apart from having to carry out the basic function of reproduction, these members also help in selecting new sites for new colonies, in addition to excavating the first galleries. The larvae and eggs are cared for by the workers who also construct and repair their nests and find food for the other insects in the colony. The soldiers normally protect the colony preventing predators from attacking.

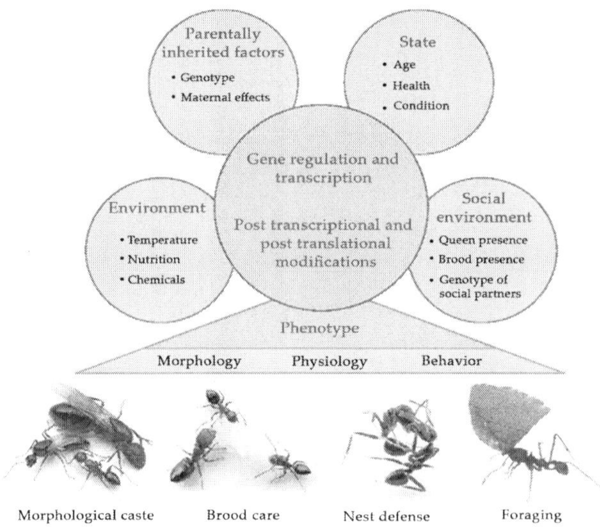

Figure 6.1: Above is an illustration of the evolution of social behavior in insects.

[Source: https://www.semanticscholar.org/paper/Genetics-and-Evolution-of-Social-Behavior-in-Weitekamp-Libbrecht/16aa5f7642393f8f3f11570f94a12aa8839bc7df]

Social insects in total normally account for 75% of the net insect biomass. It is possible to have a supercolony of interconnected nests where hundreds of millions of ants can live together. There can also be a colony of bees numbering in the tens of thousands (Cranston & Gullan, 2010).

Social behavior in insects has several advantages. Living in large colonies that are highly cooperative has proven that there is a higher amount of strength due to the numbers. In comparison to solitary insects, social insects have many more advantages. When it comes to finding food and other resources, in addition to also communicating on what they have found to other members of the colony, social insects can work together to achieve this. Living in large numbers also allows them to mount much greater defenses around their homes and resources especially in case of an attack. When it comes to competition for food and territory, social insects are much more capable when it comes to outcompeting other insects as well as some larger animals. They can quickly construct and expand the homes as required, and they can also distribute tasks in a way that makes sure everything that needs to be done is done efficiently and effectively.

When it comes to defining social insects, there are three main characteristics that they must exhibit: cooperative brood care, overlapping generations, and sterile worker class. True social insects are referred to as eusocial. Many different insects normally show social behaviors such as living in large numbers, however, this cannot be used as a measure for sociality among the insects. They have to exhibit all the stated characteristics above to be referred to as social insects (Engels, 2012).

A good example of this behavior is exhibited by termites. All termite species are eusocial insects. When you look at a termite colony, one can find individuals at the specific stages of their lifecycle. They normally have generations that are overlapping, and there is also always a constant supply of fresh adults that are ready to take charge and the responsibility of taking care of the colony. There are always cooperative efforts that are put into taking care of their young ones. There are generally 3 classs in the termite community. The king and queen normally make up the reproductive class. There are both males and females in the soldier class, and they are well adapted for the protection of the colony. In terms of the size, the solders are sterile and much larger than the other members. The third class is the worker class that includes females and males that are immature. The workers have the important job of brood care, construction, cleaning, and feeding. In comparison to social insects, solitary insects do not exhibit any of the behaviors that have been described above.

It is not possible to fit all insects into either eusocial or solitary categories because some are neither. Several different degrees can be found between eusocial and solitary:

- *Sub-social insects* - He is found just above solitary insects. These insects normally take care of their young ones for a certain time period. They normally take care of the eggs before they hatch and sometimes they may even stay with their young larvae or nymphs for a limited time period. Most of these insects normally do not utilize nests for sheltering their young, however, this rule has an exception. One species of insects that fall into this group are giant water bugs. For these insects, the female normally lays eggs on the back of the male, and the male is given the responsibility of taking care of the eggs until they hatch.

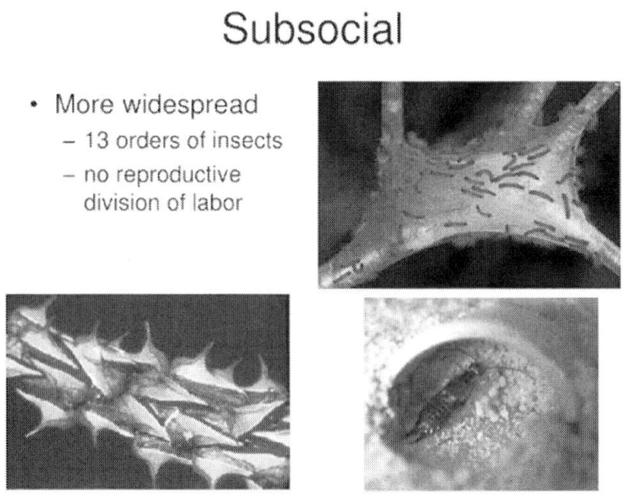

Figure 6.2: Sub-social insects do not divide the reproductive labor. They also only take care of their young for a short time period.

[Source: https://www.slideserve.com/talmai/social-insects]

- *Communal Insects* - in this group, the insects normally share the site of the nest with other insects that are in the same generation. This specific behavior can be seen in a certain lifecycle stage such as some moths in their larval stage. Insects in this group normally utilize forms of communication that are sophisticated and this also helps in providing an advantage from communal living. Some of the advantages that these insects enjoy are: being able to efficiently find and use resources, thermoregulation, and protection from predators. These insects, however, do not share in caring for their young ones. For example, the eastern tent caterpillars normally construct communal skill tents that help in providing all of them with shelter. They create chemical trails that help in sharing information about food sources which allows the siblings to track and follow this sent to the location of the food.

- *Quasi-social insects* – these insects normally exhibit a form of social behavior that is slightly more advanced. They normally cooperatively care for their offspring. These insects normally share common nests among single generations. An example of insects in this group is orchard bees. They normally have many females sharing one nest and taking care of their offspring together. Even though, all the bees share in taking care of their young, not all of them are able to lay eggs in the nest cells.
- *Semi-social insects* – these insects also share the duties of taking care of their young with other members of the same generation all in one nest. Just like in some of the true social insects, some of the insects in this group are non-reproductive workers. In this group ,however, generations normally leave before the new generations are made. The newer adults normally construct newer nests for their young. For example, during the spring, paper wasps are semi-social and have workers that are non-reproductive helping with the expansion of nests as well as brooding the new colony.
- *Primitively Eusocial Insects* – the main difference between this group and eusocial insects is in the sterile worker class. In this group of insects, the queens and the workers look the same, having very little or no differences morphologically among the class. An example of insects in this group is sweat bees. Bumblebees can also fall in this group, however, in their situation, the queen is a bit bigger than the workers, and it is much easier to differentiate the class.

6.2 TERMITES

Termites are classified under the order Isoptera in the taxonomic rank. They are eusocial insects. Initially, they were classified in an order that was separate from cockroaches, however, studies today show that termites actually did evolve from cockroaches. Entomologists have been able to describe around 3,106 species, and there are still hundreds that have not yet been described. Although termites are commonly referred to as white ants, they are neither ants nor are they closely related to ants. They normally look like ants, however, ants have much narrower waists, as well as antennae that are elbowed, and termites have thicker antennae and waists that look like a string of beads. Termites normally divide as soldiers and workers that are normally sterile. Within the termite colonies, the fertile males are referred to as kings and the colony also has wither one or more females that are fertile, referred to as queens (Adkins-Regan, 2013).

Figure 6.3: Environmental changes can greatly impact the population of termites. They mostly use feces to create their nests.

[Source: https://www.westernexterminator.com/blog/the-facts-about-using-orange-oil-for-termites/]

When it comes to feeding, they normally feed on cellulose and dead plant material, mostly in the form of animal dung, soil, leaf litter, and wood. In the tropical and subtropical regions, they are detritivores, and the manner in which they recycle plant matter and wood is considered very important ecologically. In terms of success, termites are considered the most successful insects, being able to inhabit almost all landmasses excluding Antarctica. They have many different colony sizes, ranging from hundreds to millions in population. The longest lifespan of all insects globally is that of the queen, where some of the termite queens have lived for as long as thirty to fifty years. In comparison to ants that go through a complete metamorphosis, termites normally undergo incomplete metamorphosis: egg, nymph, and adult stages. Colonies are normally referred to as super-organisms because they are part of an entity that is self-regulating (Adkins-Regan, 2013).

In the diet of some human cultures, termites are a delicacy, and in some places, they are also used as traditional medicine. Some species are considered as pests and these can damage the environment economically, by destroying plantation forests, crops, and sometimes even buildings. One of the invasive species is the West Indian drywood termite.

When termites feed on lignin and cellulose that is found on the cell walls of plants, they are not able to directly digest the cellulose, and that means they rely on protozoa and symbiotic bacteria that live in their intestines which help in supplying the majority of the enzymes that are required for the digestion of cellulose.

Termites normally cannot see, and therefore have to communicate mainly through pheromonal, mechanical, and chemical cues. These communication methods are utilized for many different reasons including defending nests, locating and fighting enemies, nuptial flight, recognition of nest-mates, construction of nests, locating reproductive, and foraging. The main way in which they communicate is through the use of antennae. Entomologists have been able to describe some of the pheromones that they normally utilize, such as contact pheromones which are normally released when workers are groomed or engaged in trophallaxis. Others are sex, trail and alarm pheromones. The frontal gland normally releases the alarm pheromone and other chemicals that are used for defense. The sternal gland normally secretes the trail pheromones and the tergal and sternal glands secrete sex pheromones (Adkins-Regan, 2013).

Figure 6.4: Above are the different termites found in a colony.

[Source: https://johnbealrealestate.com.au/how-to-tell-if-termites-are-active-in-your-house/]

Termites normally use vegetation to forage in columns on their path when they go out to look for food. One can use runways that are covered using objects and fecal deposits to identify their trail. Workers normally release pheromones on the trails which their nest-mates can use to follow them. They can detect the pheromones through their olfactory receptors. They are also able to use physical contact, vibrations, and mechanical cues to communicate. They use these signals for communicating, in case of need to cause alarm, and also for finding and making an evaluation of food sources (Berenbaum, 2010).

When they are constructing their nests, they communicate using indirect communication. They also do not give one single termite the responsibility of taking over a construction project. When termites are alone, they tend to react instead of think, however, when they are in a group, they normally have a certain level of collective cognition. Termites normally start building when they are around certain objects and structures, such as soil pellets or pillars. They normally use these objects as additions to structures that already exist, thus encouraging a building behavior in other workers. This ends up in a process that is self-organized where the information that helps in directing the activity of termites normally comes from environmental changes instead of direct contact among them.

Termites are able to know the difference between their nest-mates and those that are not from their nest through chemical communication and other chemicals they release from the cuticle and have hydrocarbons called guy symbionts, allowing them to recognize alien termite species. None of the colonies smell the same. They each have a distinct odor. This distinction comes from environmental and genetic factors such as the composition of bacteria that they have in their intestines, and their diet (Price, 2003).

6.2.1 Termite Colonies

Considering the fact that termites are social insects, they have members in the colony that have different responsibilities in accordance with their class: soldiers, workers and reproductive. It is important to understand the manner in which termites reproduce to understand this. Generally, their colonies normally have one pair of reproductive: the queen and the king. These two normally release pheromones that help in signaling mating and nesting. These pheromones normally spread all through the colony to make sure that the worker termites do not become reproductive.

The queen normally has the responsibility of starting the colony and they have to lay all the eggs up to the point in which the colony arrives at a specific size. The queen then allows for some of the termites that are immature to undergo development into secondary queens which help in laying the extra eggs that are necessary for ensuring the colony continues to thrive. The queen and king normally live for quite a long time: 15 to 30 years. If there is a situation where the king or queen dies, or the colony has greatly increased in size, it leads to a change in the type and concentration levels of the pheromones. This then leads to either one or more workers molting into termites that are reproductive or alates (Price, 2003).

Subterranean termite colonies normally take about 6 to 7 years to mature, and in this time period, there is a production of the alates. Depending on the species, the alates can either be pale reddish-brown to brown-black. During the early spring or late winter, female and male alates normally get out of their nests to go on a mating session with hundreds to millions of winged termites. The alates do not fly too well, so they are only able to fly a short distance from their older nests. It is however possible for the alates to be swept to longer distances by the wind. Most of them are normally eaten by spiders, birds, lizards or frogs during this period of mating, so there are not many alates that survive to locate new sites for nesting, undergo reproduction, and create new colonies. This means that the colonies have to produce many alates, to ensure that some survive the mating period (Price, 2003).

Figure 6.5: In termite colonies, different class have different jobs. They work in a cooperative manner.

[Source: https://www.terminix.com/termite-control/colonies/]

During this mating period, female and male swarmers normally couple up. Pheromones are released by the female alate for the attraction of males so that they can nest and mate. Shortly after landing, the winged termites normally drop their wings and live the rest of their time without wings. The nesting site that the female and male couple look for is normally near moist soil.

Before they mate, the new queen and king burrow into the ground and dig a chamber that they consider royal, in the moist soil. They then use soil, their waste and saliva to seal their nests. After the king and queen mate, they spend the rest of their lives in their royal chamber underground and the worker termites feed them. Within a few days or weeks, the queen then ends up laying the eggs (6 to 12), all of which will contribute to the establishment of the new termite colony. A fully mature queen is able to lay thousands of eggs every year. As there continues to be an increase in the size of the colony, the queen has the ability to increase the number of reproductive, so as to help in the production of enough workers for feeding the colony. A single mature colony may have around one million termites (Gordon, 2010).

- The Workers: Workers are soft-bodied and wingless in most termite colonies. They account for the highest number of termites in most colonies. These are the termites that are mostly seen in wood that is termite infested. The sterile adults and immature termites are the workers in some termite colonies. These normally take care of the immature termites and the eggs, they take care of the building, maintain the structures of the colony, feed and groom nestmates, and forage for food.
- The Soldiers: These members normally work to protect the colony. They look much different from the other members of the colony and they are also slightly larger than the workers. In comparison to the workers, they also have larger mandibles and darker heads. They defend and guard the colony against enemies.
- The Reproducers: These are the ones that procreate. Their bodies are much darker. They look very different from the other members of the colony. As had been stated before, the reproducers normally have two pairs of wings during maturity. These are referred to as alates. They shed their wings after the swarming flight and this is when they start establishing a new colony.
- The Queen: This is the largest termite physically, having a mass that is seven times larger than the workers. Her only job is to lay

eggs. Other members can replace the queen and king in case they died.

- New Branch: It is possible to have the division of a colony that already exists, thus establishing a Subterranean termite colony. This colony is referred to as a sub-colony and it is able to produce its own reproductive. It can even rapidly develop into an independent colony.

6.2.2 Types of Termites

Subterranean termites: These are normally found in hotter climates. They have at least six different species: the desert subterranean termite, arid land subterranean termite, Western subterranean termite, Formosan subterranean termite, Light Southeastern subterranean termite, Dark Southeastern subterranean termite, and the eastern subterranean termite. This genus can create large colonies. They normally rapidly eat through wood, and this makes them the most destructive termite type. Of the species, the Formosan termite has the largest colony size, making it the most destructive of the species. The Formosan termite has a darker yellow-brown color, making it easier to distinguish from other subterranean termites.

Figure 6.6: In the United States, subterranean termites are the most destructive termites.

[Source: https://content.ces.ncsu.edu/termites-biology-and-control]

Drywood termites: There are four main types of drywood termites: the desert drywood termite, the western drywood termite, the West Indian drywood termite, and the Southeastern drywood termite. These normally tend to attack furniture, flooring, frames, and structures. Unlike other termites, they receive all their nutrition from wood and they do not need to nest in moist soil. These termites are able to create many different colonies, making them the hardest colony to treat. They survive better in cooler temperatures.

Dampwood termites: Just like the drywood termites, they are able to survive without nesting in wet soil. These normally look for wood that is decaying, woodpiles and places that have leaks. They have red-brown heads and they are much bigger than the other termite types. They hardly invade homes due to their need for a free water source.

6.3 ANTS

Ants, come from the family Formicidae, and they are eusocial insects. Together with bees and wasps, they are part of the order Hymenoptera. Ants have undergone an evolution from vespoid wasp ancestors and then, later on, became diverse after the flowering plants rose. There has been a classification of more than 12,600 species of ants of an approximate 23,000 species. It is easy to identify an ant due to its distinct node-like structure leading to its slim waist and their elbowed antennae (Adkins-Regan, 2013).

Ant colonies come in various sizes, from a couple of dozen predatory ants cohabiting in cavities that are small and natural, to colonies that are highly organized and live in large territories consisting of several class of soldiers (dinergates), as well as wingless, sterile females, majority of them being workers (ergates), and other specialized groups. The majority of ant colonies also comprise one or more fertile females, referred to as queens (gynes), and some fertile males, referred to as drones (aner). The term super-organisms is used to refer to these colonies because they can operate as an entity that is fully unified, working collectively to make sure that the colony is well supported.

Almost every landmass on Earth is colonized by ants. Only Antarctica and a few inhospitable or remote islands lack indigenous ants. They can thrive in the majority of the ecosystems and may end up forming 16-26% of the terrestrial animal biomass. They have been able to succeed in all these environments due to their high-level of social organization, as well as their ability to defend themselves, tap resources, and modify habitats. They have also been able to co-evolve with other species, leading to mutualistic, parasitic, commensal, and mimetic relationships (Adkins-Regan, 2013).

Figure 6.7: Ants have shown mannerisms of coexisting with other species, in addition to being highly social and cooperative.

[Source: https://www.popsci.com/six-amazing-facts-about-ants/]

Societies of ants have an ability to solve problems that are complex, communicate between the individuals, and divide labor. This is parallel with human societies and thus has been a subject of study and inspiration. Many different cultures in the human population make use of ants in rituals, medication, and cuisine. There are even some species that are used due to their ability to act as biological pest control agents. They are able to exploit resources, and this can lead to a conflict with humans because they can invade buildings and damage crops. One of the invasive species is the red imported fire ant, and they can establish themselves in locations that they may have been introduced accidentally.

6.3.1 Ecology and Behavior

Communication: Ants use touch, sounds and pheromones to communicate with each other. Ants have a much more developed use of pheromones, for example, the red harvester ant, in comparison to other hymenopteran groups. Similar to other insects, ants use their mobile, long, and thin antennae. These antennae are paired, and they help in providing information on the intensity and direction of scents.

Since, most ants live on the ground, the soil surface is utilized as a means of leaving pheromone trails that other ants can use to follow. For ant species that normally do foraging in groups, when one of the foragers find food, they can mark a trail when going back to the colony; other ants are then able to follow the trail, and as they head back to the colony with the food, they reinforce the trail. When the source of food is exhausted, the ants that are returning to the colony do not mark any new trails, and this scent then ends slowly. This is a behavior that allows ants to adapt to environmental changes. For example, when an obstacle blocks a path that that has been created to a food source, the foragers then abandon that path and start looking for new routes. When the forager successfully finds a new food source, a new trail is left as it takes a shorter route. More ants follow the successful trails and reinforce the much better routes,therefore, leading to a gradual identification of the best path (Price, 2003).

There are many other uses of pheromones in addition to creating trails. When an ant is crushed, it releases an alarm pheromone that ends up sending ants that are nearby into an attack frenzy, therefore, attracting more ants that are far away. There are some species that normally use the propaganda pheromone for the confusion of enemy ants making them fight amongst themselves. They are many different structures that produce pheromones including the hind tibia, sternum, rectum, pygidium, glands on the hind-gut and poison glands, and Dufour's glands. Pheromones are also passed by trophallaxis, mixed with food, and exchanged, leading to the transference of information within the colony. This makes it possible for other ants to know the task group that a different ant colony member belongs to, for example, nest maintenance or foraging. In species of ants that normally have a queen class, immediately a queen stops the production of a certain pheromone, the worker ants start to raise a new queen in the colony. Some ants use stridulation to produce sounds with the use of their mandibles and gaster segments (Price, 2003).

Defense: To protect and attack themselves, ants bite, and in some other species sting, many a time by spraying or injecting chemicals such as protein components, piperidine and alkaloids in fire ants, and formic acid in formicine ants. In South America and Central America, bullet ants are considered to have the most painful sting amongst all insects, however, this sting is not fatal. On the Schmidt sting pain index, this sting is given the highest rating. The jack jumper ants have a sting that can be fatal and there has even been the development of an anti-venom.

Figure 6.8: The bodies of ants undergo division into 3 main parts: abdomen, thorax and head. On the different sides of the ants' mouth are the mandibles which are utilized for fighting, digging and carrying food.

[Source: https://sciencing.com/do-ants-protect-themselves-8709450.html]

Ants not only defend themselves from predators, however, they need to also protect themselves from pathogens. The hygiene of the colony is sometimes maintained by some worker ants, and they do this by disposing of dead nestmates through undertaking or necrophory. Dead Atta Mexicana ants release Oleic acid, and this has proven to trigger the necrophoric behavior. *Linepithema humile* ants trigger this behavior by reacting to the absence of certain chemicals (iridomyrmecin and dolichodial) that are found on the cuticles of the nestmates that are alive (Gordon, 2010).

It is possible to protect nests from certain physical threats such as overheating and flooding by elaborate nest architecture. In the hollows, there is an arboreal species called *Cataulacus muticus*, and the workers normally deal with flooding by having to drink the water that is inside the nest and getting rid of it by excreting it outside. Within the cavities of wood in mangrove habitats, the *Camponotus anderseni* ants nest, and thy deal with submergence underwater through making a switch to an aerobic respiration.

Learning: Many different animals imitate to learn behavior, however, with ants, there is the use interactive teaching, and these are the only group of animals, apart from mammals, that can achieve this. With the *Temnothorax albipennis* species, a knowledgeable forager can lead a nest that is naïve to food that is newly discovered through the process of tandem running. These ants that are following are then able to attain knowledge from the leading ant. The leader has a level of sensitivity to the process, slowing down when the followers are lagging and speeding up when they are getting too close (Engels, 2012).

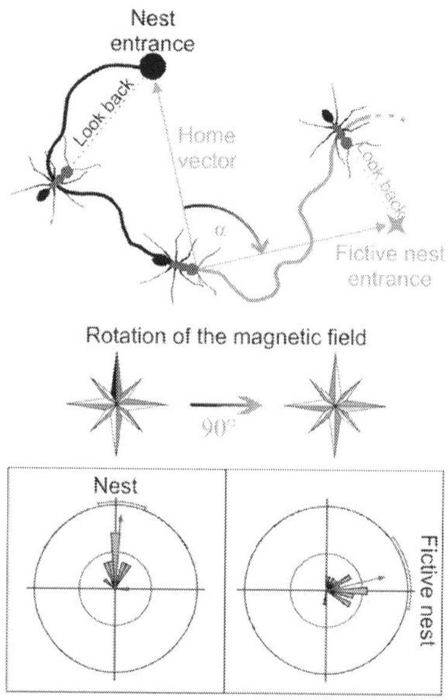

Figure 6.9: Ants learn to navigate in many different ways. For example, the Desert ants utilize a very important tool for navigation for getting back to where their nests are, even in terrains that are very unfamiliar.

[Source: https://www.degruyter.com/view/journals/nf/25/2/article-p109.xml]

Nest construction: Many different ant species build complex nests, however, the majority of the species are nomadic and therefore their structures are not permanent. It is possible for ants to either build the nests on trees or from subterranean nests. These nests may be found inside acorns, hollow stems, logs, under logs or stones, or even in the ground. The carefully select plant matter and soil to construct the nests. Some species, such as *Temnothorax albipennis* end up avoiding locations that have dead ants because this can be a sign of disease or pests. At the first sign of threats, these species are quick to abandon nests that they have already established (Engels, 2012).

Some of the species, for example, those that utilize natural cavities, can be opportunistic, allowing them to utilize controlled micro-climates that they are able to get inside human homes and also other structures that are artificial to construct nests and house the colonies.

Food cultivation: In general most ants are indirect herbivores, scavengers, and predators, however, there are a few that have been able to evolve ways in obtaining nutrition that is specialized. There is a belief that majority of the ant species that are indirect herbivores depend on specialized symbiosis with the microbes in the gut for upgrading the nutrient value of the food they collect, allowing them to survive in regions that are poor in nitrogen levels, such as canopies in the rainforest. Leaf cutter ants exclusively feed on fungi that they collect and grow within their colonies. They collect leaves in a continuous manner which they take to the colony, and thereafter cut them up into smaller pieces and place them in fungal gardens (Gordon, 2010).

The Ergates specialize normally assign tasks in accordance to their sizes. The large ants are assigned the work of cutting stalks, the small ants have the work of chewing the leaves and the smallest ants have the work of tending to the fungus. Leaf cutter ants have a level of sensitivity that allows them to recognize when the fungus is reacting to different plant materials, which means that they detect chemical signals that they receive from the fungus. If the fungus finds a certain type of leaf to be toxic, the ants get a sign and they no longer collect that specific leaf. The fungi produce structures called gongylodia that the ants feed on. The exterior surface of the ants has symbiotic bacteria that allow the ants to produce antibiotics that work by killing bacteria that may enter the nest and harm the fungus (Gordon, 2010).

Navigation: Ants that forage end up traveling distances of up to 700 feet from their nests, but the pheromones they release on the trail allow them to get back to their nests no matter how dark it is. In arid and hot regions, ants that forage in the day may die through desiccation, and therefore, they are required to discover shorter routes back to the nest to lower that level of risk. The genus Cataglyphis has diurnal desert ants, for example, the Sahara Desert ant, which can navigate by ensuring that the direction and distance traveled is kept track of. They can keep track of the distance that has been traveled through an internal pedometer that allows them to know the number of steps they have taken, in addition to evaluating the optical flow; the movement of objects that they see in the visual field. They use the position of the sun to measure the directions. They then integrate this data to locate a short route that can take them back to their nests. They are also able to utilize available visual landmarks, in addition to tactile and olfactory cues, for navigation (Cranston & Gullan, 2010).

Ants have compound eyes that have specialized cells allowing them to detect polarized sunlight, which they use for the determination of direction. These detectors of polarization have a certain level of sensitivity in regions of the light spectrum that are ultraviolet. With the army ant species, when there is a separation between a group of foragers from the main column, the ants may end up turning back and forming a circular ant hill. This makes the workers run around the circular ant mill in a continuous manner which ends up killing them due to exhaustion (Gordon, 2010).

Competition and cooperation: Societies are not the same in all and species. The biggest and most basal ants are the Australian bulldog ants. In comparison to other ant species, in addition to being eusocial, they have poorly developed social behavior. Each of the ants has to hunt alone, with the use of their large eyes, instead of chemical senses, for finding prey.

Some species of ants thrive by attacking and taking over neighboring ant colonies. Others are just as aggressive but less expansionist; they thrive by invading colonies mainly to steal larvae or eggs, which they either raise as slaves or workers or eat. Some extreme specialists within these ants are slave-raiding, such as amazon ants, are not able to feed themselves and they require captured workers for their survival (Gordon, 2010).

Figure 6.10: Ants have been studied for their highly famous collective intelligence. Scientists have been able to start many projects that show the importance of linking natural behavior and ecology with certain aspects of neuroscience.

[Source: https://phys.org/news/2019-02-ant-cooperation-revealing-brains.amp]

Ants can identify their nestmates and kins through their scent, which is derived from secretions that are laced with hydrocarbons, coating their exoskeletons. If there is a separation between an ant and its colony, the scent of the colony is lost. When an ant joins a colony without the scent of the colony, the ants attack it. In addition, the main reason why ants from two different colonies would attack each other, although they are the same species, is because they have different genes responsible for the production of pheromones. This characteristic is however not seen in the Argentine ant, because they lack genetic diversity, and due to this, this species has become a global pest (Klowden, 2013).

Some ant species are parasitic, and they join host ant colonies, establishing themselves as social parasites. These ants are fully parasitic and they do not have workers, and they, therefore, depend on the food collected by their hosts. Many ant genera have this form of parasitism, however, the parasitic Ant has to be a closely related species to that of the host. To enter the nest of the host ant, many different methods are employed. Before the hatching of the first brood, the parasitic queen can enter the host nest and establish herself before the colony scent is developed. Some parasitic ant species use pheromones that can trick the host ants into carrying the parasitic queen into their nest or confuse the host ants in general.

Some species of ants end up having a conflict between the sexes of the species, with the reproducers competing for the production of offspring that have the closest relation to the individual reproducers. There has been an extreme form that involves producing clonal offspring. *Wasmannia auropinctata* shows an extreme sexual conflict, where the queen uses a process referred to as thelytokous parthenogenesis to produce diploid daughters, and the male uses a process that allows the maternal contribution to be removed from the diploid eggs, thus producing haploid males that are the father's clones (Klowden, 2013).

6.4 BEES

Bees live together in large, well-organized family groups, making them social insects. They normally engage in many different complex tasks that many solitary insects do not practice. Bees have developed certain behaviors that allow them to successfully thrive in social colonies, including division of labor, defense, environmental control, complex nest construction and communication, making bees together with other creatures on earth very fascinating.

There are three kinds of adult bees in a bee colony: a queen, drones, and workers. Thousands of worker bees work together to rear broods, collect food and build the nests. Each of the members has a specific role they are supposed to perform in relation to the age of the adult. However, to survive and reproduce requires the combined efforts of all the members in the colony. The individual bees (queen, drones, and workers) can't survive on their own. They require the support of the whole colony (Wigglesworth, 2012).

Figure 6.11: There are various species of bees that end up dying after they sting due to the fact that they have stingers attached to their abdomen, and these stingers have hooks or barbs on them. This means that a part of their abdomen is ripped away after they sting.

[Source: https://www.unenvironment.org/news-and-stories/story/celebrating-greatest-all-pollinators-bees]

A colony comprises of one queen, thousands of worker adults, and several hundred drones during summer and late spring. The presence of the workers and the queen helps in maintaining the social structure of the colony, together with an effective communication system. The activities necessary for the survival of the colony are controlled through communicative "dances", and distribution of chemical pheromones within the individuals. The distribution of labor among the bees is all dependent on the bee's age, however, these needs vary with the colony's needs. The strength of the colony and reproduction depends on the queen, the worker force size, and the number of food stores. With an increase in the size of the workers in the colony by approximately 60,000 workers, the colony's efficiency also increases (Wigglesworth, 2012).

6.4.1 Queen

For bees, there is only one queen in each colony, and this rule is only different in the situation of a varying period after supersedure or swarming preparations. Since the queen is the only female in the colony that is sexually developed, she primarily functions as the reproducer. She can produce both unfertilized and fertilized eggs. She lays the largest number of eggs in the early summer and spring. During this period of peak production, she can lay up to 1,600 eggs in a day. In early October, the queen gradually stops laying eggs and lays few to no eggs up to January (the next spring). In a year, one queen has the ability to produce up to 250,000 eggs, and in her lifetime, possibly more than a million (Price, 2003).

It is quite easy to distinguish a queen from the other members of the colony. She has a much longer body in comparison to the workers and the drones, especially during the period in which she is laying eggs and her abdomen has become greatly elongated. Approximately two-thirds of her abdomen is covered by her wings, in comparison to the drones' and the workers' wings which when folded, are as far as the tip of the abdomen.

The thorax of the queen is also a bit bigger than that of the worker, and she also lacks functional wax glands and pollen baskets. In comparison to the worker, she has a curved stinger which is also longer than that of the worker, however, it has shorter and fewer barbs. Queens can live for as long as 5 years, however, on average they live 2-3 years (Price, 2003).

Queens have another major function in the colony, involving the production of pheromones that help in acting as a social "glue", providing the individual identity to the colony, in addition to unifying it. The main pheromone she releases is 'the queen substance', which she produces with the help of her mandibular glands. The chemical production and egg-laying capabilities of the queen are dependents of the colony's qualities. The temperament, size. and quality of the colony are dependent on the queen's genetic makeup, as well as that of the mating drones.

Approximately one week after the queen emerges from the queen cell, she exits the hive with the purpose of mating with several drones that are in flight. Because she has to fly a distance away from her colony to mate, she has to go round her hive so that she can remember its location when she is coming back. She flys away from her hive to prevent inbreeding. She exits the hive on her own and is normally gone for approximately 12 minutes. The afternoon is the best time for the queen to mate, which happens with 7-15 drones. For this event to occur, they have to be at an altitude greater than 20 feet. The queen's pheromones allow the drones to locate and recognize her. If the queen's mating flight is delayed by bad weather for more than 20 days, she ends up losing her ability to mate, meaning she will only produce unfertilized eggs, resulting in drones (Price, 2003).

After she mates, she goes back to her hive and starts the process of laying eggs in approximately 48 hours. Each time she lays an egg, she releases a number of sperms from the spermatheca. These eggs are destined to be either a queen or worker. However, she does not release sperm if she lays her egg in a larger drone-sized cell. The colony's worker bees constantly feed the queen royal jelly. The quantity of eggs laid by the queen is highly dependent on the amount of food she is fed, as well as the size of the workers that can prepare beeswax cells for the eggs, as well as taking care of the larva hatching from the eggs in approximately three days.

When there is not an adequate amount of queen substance being secreted, the workers start making preparations to replace her. Following supersedure, the old queen and the new daughter are allowed to remain in the hive for a while. The queen that will then take over is developed from either young worker larva that is not older than 3 days, or from fertilized eggs. There are three different circumstances in which the new queens are raised: swarming, supersedure, or emergency. If by any chance the queen is accidentally removed, lost or killed, the other individuals in the colony start selecting younger worker larvae for the production of emergency queens. If an old queen starts failing (lower queen substance production), preparation start for raising a new queen. When queens are produced due to supersedure, they tend to do a better job than emergency queen cells, because they are able to receive much more food (royal jelly) as they develop. When you compare, those that are produced while preparing for swarming develop within the brood area (Price, 2003).

6.4.2 Drones

These are male bees, and in the colony, they are the largest bees. Drones are only found during summer and late spring. In comparison to the worker or the queen, the drone's head is much bigger, and its compound eyes end up meeting at the top of its head. They do not have stingers, wax glands, or pollen baskets. They have the purpose of fertilizing the virgin queen during the mating flight period. A week after they emerge, they become sexually mature, however, they die immediately after mating (Shields, 2017).

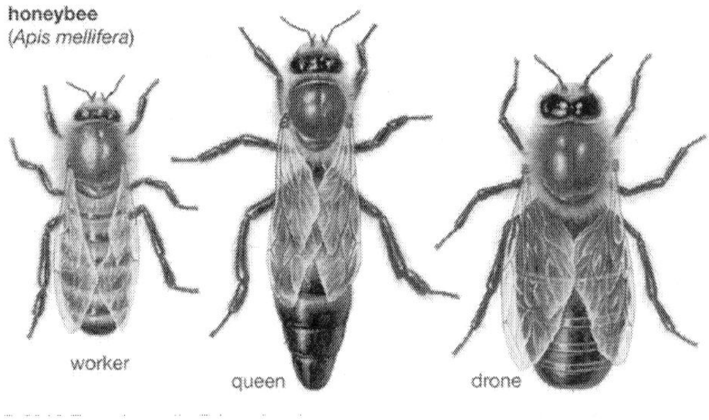

Figure 6.12: This is the difference between the different bee members. The different individuals have different duties and this has proven to be very efficient in maintaining beehives or societies.

[Source: https://www.britannica.com/animal/drone-bee]

Even though drones have no main purpose or work in the hive, they are still very important for the colony to function normally. While they normally depend on workers to give them food, they can feed themselves inside the hive, after they are four days old. Because they eat approximately 3 times more food than workers, when they are high in number, they can end up placing a certain level of stress on the food supply of the colony. They normally live in the hive up to eight days old, where they start taking orientation flights. They normally fly -aways from the hive between midday and 4 PM. It has never been confirmed if drones can feed on flowers because no observations have been made. In the fall, when cold weather begins, and there is a scarcity of nectar/pollen resources, the drones are forced to leave the hive and they are left to starve in the cold. However, when colonies lack a queen, the drones can be allowed to stay in the hive for an indefinite time period (Shields, 2017).

6.4.3 Workers

These are the smallest in the colony, however, they account for the majority. Workers are sexually underdeveloped females, and under normal conditions within the hive, they are not able to lay eggs. They have structures that are much more specialized, such as pollen baskets, wax glands, scent glands, and brood food glands, all of which make it possible for them to do all the hive's labors. They can ventilate and air-condition the hive, guard the entrance, build beeswax combs, handle incoming nectar, remove debris, care for the queen, feed the brood, and polish and clean the cells, all this in their first few weeks as adults. When they grow and become field bees, they forage for plant sap (propolis), water, pollen, and nectar.

The worker bees have a life span of six weeks during the summer. Rearing of workers in the fall allows them to live for as long as six months, providing the colony with the opportunity of surviving the winter, and helping in the production of newer generations in the spring before dying (Engels, 2012).

6.4.4 Laying Workers

When the queen is not available in a colony, several workers develop their ovaries to allow them to lay unfertilized eggs. Normally, in the presence of a queen, the brood, and the queen's pheromones work to inhibit the development of the workers' ovaries. When there are laying workers in a colony, it means that the colony has not had a queen for more than one week. It is however possible to find laying workers in a normal colony, either when the colony has a poor queen or during the swarming season. It is easy to recognize a colony that has laying workers: they rare small- bodied drones in worker sized cells, and there is an approximately 5-15 eggs per cell.

Besides, these bees randomly scatter their eggs in the brood combs, and this makes it possible to find the eggs on the cell sides rather than the base. These eggs may end up not hatching, and the majority of these that don't hatch die before they are mature.

6.5 WASPS

Only certain members of the family Vespidae exhibit Eusocial behavior among wasps. These wasps are generally referred to as yellowjackets, hornets, and paper wasps. They mix wood fibers with saliva to create a paper-like material, and this can be used to construct brood cells as well as other components of the nest, thus building communal nests. The cells in which larvae are reared (brood comb) are normally constructed to look like an upside -down umbrella with the open parts of the cells towards the ground. The worker wasps normally cling to the bottom of the comb as they perform different housekeeping chores including feeding the larvae and guarding the nest. The majority of the wasps in a single colony are workers. The nests they construct have many cell tiers which are covered with a shell outside, having one opening at the bottom (Engels, 2012).

All of the social wasps are carnivorous; they feed on flies and caterpillars. The worker wasps capture their victims and chew them up into a paste that can easily be fed to the larvae, and the larvae then create a syrup that is nutritious for the adult wasps to feed on. A tiny colony of Yellowjackets (approximately 200), is able to prey on 6000 caterpillars in one summer (Engels, 2012).

Figure 6.13: wasps that are social normally construct their nests from wood fibers which they pick up and mix with water to create a form of paper maché which is able to produce structures that are very strong and long -lasting.

[Source: https://www.bbc.com/news/amp/science-environment-41042948]

The structure of the class in the social system of the wasps is mainly maintained through interactions that are considered aggressive, among the members of the colony. If the queen is not able to properly sustain her role, another fertile female replaces her, and this replacement gets the responsibility of producing eggs. Unmated workers lay unfertilized eggs from which the eggs develop.

Social wasps are active all year in tropical climates. They form new colonies during the swarming period, where the fertile queen leaves the parent nest together with a group of workers, and they separately set up for themselves housekeeping. Wasp colonies are set up in the early spring in a temperate climate, by one or more queens who were able to mate in the summer before, and went into hibernation all through winter. The queen builds a small nest that only has a few brood cells. After she lays a few eggs, she takes care of and feeds her offspring up to the time they become adults. These adults are all female and end up as her workers. They deal with housekeeping duties, foraging, and brood care. The queen is then able to proceed with egg -laying, growing the colony size all through the summer. The structure of the colony starts to break down in early fall. The eggs that are unfertilized produce males who end up mating with the females that newly emerging, and these females will go into hibernation during winter and create new colonies the next year, and the cycle continues (Engels, 2012).

6.5.1 Types of Social Wasps

In North America, there are approximately 20 species of social wasps and greater than 700 species globally. These can be divided into 3 groups:

- *Yellowjackets* – these wasps build their nests in cavities underground, for example, rodent burrows. They are roughly the size of honey bees, however, they are black and have yellow marks on their body. They are similar in a way to hornets because they build paper nests. It is common to find Yellowjackets hovering around, to and from the opening of the nest, or around areas where they look for food, such as around garbage cans (flies). For Yellowjackets, the nest can have a maximum of 5000 workers at a time, and the majority of these workers hardly travel greater than a few meters from the nest when foraging. They have a sting that is quite painful, however, wasps are hardly ever aggressive, unless attacked or disturbed at their nest.

Figure 6.14: Above shows a clear visual distinction between paper wasps and yellow-jacket wasps. They have many differences from their nests to their appearances and there are many different ways in which you can differentiate them.

[Source: https://www.rescue.com/latest-buzz/outdoor-pests/how-to-tell-a-wasp-from-a-yellowjacket/]

- *Hornets* - Nests that are created by hornets are normally in areas that are wooded above ground, on the branch of a tree, however, they may still be constructed on a shrub, house siding and utility pole. The workers found in each nest are dark and have cream, light yellow or white -colored markings on their face, thorax, and abdomen. They are approximately one-inch long. Hornets can be dangerous and sting as a colony, but only when the nest is threatened or disturbed. Their nests normally have a paper wall surrounded that they build to protect them from natural enemies and the weather.
- *Paper wasps* – generally, we find these nests below shelter overhangs, where they gain protection from the rain and winds. These wasps normally construct their nests, which are also umbrella -shaped, all hanging by a short stalk on the bottom side of a building eave. They however only construct a single tier of cells, and they have no paper walls covering the outside of their nests. Their colonies normally have less than 25 wasps, however, later these numbers may go up to 100. In comparison to the other social wasps, these are a bit longer and much slenderer, and their colors vary with the many different species.

Social wasp colonies only last for one year. All these types of wasps always construct new nests from the beginning every single year, and they never use the previous nest again. Only the fertilized queens overwinter. All the workers end up dying with the first frosts.

6.6 SOCIAL INSECTS INSPIRE HUMAN DESIGN

Certain events that could either be natural disasters, accidents, or targeted or random attacks can negatively affect the human infrastructure. Our communication networks, supply chains, and transportation networks are increasing in size, and they are also becoming much more complex with the growing population.

It is important to ensure that these networks do not fail or become vulnerable. Weirdly enough, we can get this information from social insects, seeing as they have been able to maintain large colonies effortlessly, or maybe with a lot of effort but great discipline.

When you think about it, social insects, such as wasps, bees and ants live in a common area (nests). They:

- Have generations that overlap with old and young members living together.
- Have queens (fertile females) from which all the nestmates are produced.
- May have one or more than one queens.
- Work as a group to raise mates within the nest.

It is true to say that each of the individuals in the nest act very equivalent to a neuron in a brain. They react and interact with the behavior of the neighbors that are closest and living very close to the nest. Impressive feats can be noted in this social behavior, for example, termites that can build mounds that are quite large or ants that work together to construct rafts or bridges. The nest is also provided with a certain level of resilience when they have to deal with an attack. They end up having to work together, and in a situation where a small failure occurs, it can be fixed before it becomes a huge problem. This is referred to as collective behavior. Many species exhibit this behavior, including humans, birds, fish and some other insects.

Parallels can be drawn between social insect societies and human societies. However, there has been a continued study dealing with collecting behavior which has had implications in robotics, physics and math. This can be very helpful in designing infrastructure that is much better and more resilient.

Figure 6.15: There are many things we as human beings can learn from social insects. They have many different building behaviors and building processes that we as humans can emulate to construct more sustainable infrastructure.

[Source: https://crawfordlab.ca/2017/05/16/how-and-why-are-humans-like-social-insects/amp/]

6.7 CONTRIBUTING FACTORS

Social insect resilience is contributed to by two main factors: job switching behavior and redundancy, and decentralized control. In social insect colonies, there is no blueprint nor leader. This is a form of control that is decentralized and this means that each of the members can easily be replaced, making it possible for colonies to build a level of resistance to losing members (Gordon, 2010).

Besides, when the colony is damaged, it can deal with this situation as it happens without the required communications, which are time- consuming, with managers. The systems that humans are using have proven to be more decentralized, mainly since they depend on the internet, even though the majority of internet services are becoming centralized.

When you think about the nature of Google, which is streamlined, not only is it able to provide a search engine, but also Google documents, a calendar, and an email service. It is much more convenient and easier to manage the online lives of humans through a centralized service. However, if there is a corruption of an aspect of Google, through an attack online, it is possible to cripple the access of the majority of these services, due to the cascading failures through the system (Gordon, 2010).

In contrast to the phrase "busy as a bee", majority of these colonies normally house many individuals in the nest who do not work. They are referred to as inactive workers or redundant individuals, and these individuals function as a safety net and thus provides the nest with an opportunity to rapidly respond to either an attack from a predator or losing workers.

In response to disruption, insects have the ability to switch jobs when responding to disruption. Inside a nest, specific members have specific jobs. This improves the level of efficiency within the nest. However, in the event of a disruption, the gaps can be filled by certain individuals switching jobs.

When we study how social insects behave, as well as the manner in which they are able to maintain resilience in times of disruption, we are able to build upon these lessons to promote the development of our infrastructure and also in protecting the infrastructure from large-scale failure.

All in all, there is a lot we can learn from social insects as human beings. We should not take these teachings for granted.

7

Insect Locomotion

CONTENTS

7.1 INTRODUCTION

There are different species of animals around the world. These animals differ in terms of their morphology as some of them have legs while some don't. Among these animals, some are constantly in motion while others stay in one place for the rest of their lives. Some move with purpose while others drift wherever the currents take them and therefore they use different forms of locomotion. Locomotion is not similar to movement and as much as all animals move not all of them locomote. In ethology in which animal behavior is studied, locomotion is defined as an animal movement that results from progression from one point to another. Therefore, animals that spend their entire life cycle at one point are referred to as sessile while those that move around are referred to as motile. Evolution of locomotion has occurred overtime to ensure that animals are successful in looking for food, reproduction, escaping predators and moving away from unfavorable habitats.

Figure 7.1: The viceroy and the monarch butterfly which always appears to be similar.

[Source: https://en.wikipedia.org/wiki/Insect]

7.1.1 Principles of Locomotion

Locomotion exists in two forms either active or passive locomotion. Passive locomotion is found in different animals such as the jellyfish and this movement type, the environment provides the transportation. In this form of movement, no effort by the muscles is required. The disadvantage is that the animal is left at the mercy of wind or waves (Herreid, 2012).

Most of the animals exhibit locomotion that is active at the different stages in their life cycle. for animals to move from place to place with purpose, they require a means that provides them with propulsion and also exercise control over their movements. In most cases, they have muscles attached to their legs to produce the force they require for their movement. The amount, type and location contractions are governed by a nervous system. The complexity of the nervous system will differ depending on the type of animals. Control of the nervous system produces movements that have the rhythm of the appendages and this results in their locomotion.

Active locomotion can either be appendicular or axial. In locomotion that is appendicular, the appendages of animals interact with the environment by pushing hence providing a force that is propulsive to facilitate movement. Axial locomotion happens when the animal needs to alter its shape to achieve motion. Whether an animal uses passive or active locomotion, the environments they occupy fall into categories that differ. The environments also require different locomotion forms and they include the fossorial environment found underground, the terrestrial environment, the aerial environment, tree -dwelling and finally the aquatic environment. All these environments are similar in terms of the restraints that exist in relation to movement or motion and these include weight or inertia, gravity and drag. Drag is any force that can restrict movement especially in fossorial locomotion and restricts motion that is forward. Animals also require muscles that are strong in order to maintain movement such as flight or movement underground against the gravity force. Similarly, muscles are also required to overcome the force of inertia in order to move forward and the more weight it has the more inertia it will have (Herreid, 2012).

Insects are the largest group of invertebrates that fall within the phylum Arthropoda. Insects possess an exoskeleton that contains chitin, a body divided into three parts, compound eyes, joint legs and a pair of antennae. Insects are a species that is diverse and include more than half of the animal species of which insects represent half of them. Insects inhabit different environments, although some insects inhabit oceans other than terrestrial environments which are a habitat that is commonly inhabited by other arthropods such as crustaceans. Insects hatch from eggs and are constrained within their exoskeletons. Their development differs depending on insect some undergo three stages while others undergo four stages.

Insects move around in ways that differ and this includes either walking, swimming, flying, jumping or even floating. Movement may also vary depending on the stage of development of the insects for example some insects may walk in their first stages of life and may grow into flying as they develop into adults. Insects can also make adaptations in terms of their movement for example in case they lose a limb they can change how they move and this mostly applies to insects that walk but it can sometimes apply to those that fly or swim. The anatomy of insects determines the nature of their movement. If insects have wings most of them fly, if they have legs they can also be able to walk and this can be an addition to other movements. Insects can also adopt a tripedal gait. This means that they move in angles that are alternative and this can be moved in the front, side or rear on one side with the middle on the other side (Rasnitsyn & Quicke, 2007).

Insects are the only insects that have evolved in terms of their flight and thy all evolve from a common ancestor. Insects also spend at least part of their life underwater. This has led to larval adaptations which include gills. Some adult insects are also aquatic and have therefore adapted to swimming. Some insect species such as water striders have also adapted to aquatic life and can walk on water surfaces.

Different insects have mastered different ways of movement for example roaches crawl, bees swarm, moths fly, mantids strike, diving beetles swim, caterpillars crawl, dragonflies dart, maggots squirm, water boatmen paddle, mole crickets burrow, mosquito larvae wriggle, fleas jump, whirligigs spin, collembolans spring, water striders skate, army ants march and backswimmers dive. The difference between insects and other forms of life is their independence in movement and movement that is goal- oriented.

7.2 LEGS AND LOCOMOTION

Insects have three pairs of legs and there is a pair for each segment. The legs of insects move a set on intrinsic and extrinsic muscles which is also made by a set of sensory receptors as well as with the help of structures that are mechanical such as their hairs, spines and pads that are adhesive. Most insect's use their legs primarily for terrestrial movement but modifications on their legs have also allowed them to be used in other different types of locomotion which include jumping, swimming or also movement on the water. The legs can also be modified for other purposes or functions which include grasping, grooming; a collection of pollen grains and production of silk (Wigglesworth, 2012).

The exoskeleton of an insect can be a burden to small- sized animals and this is made easier for insects by the positioning of their legs. This also improves the stability of insects. The legs of insects therefore also function to provide support to its body in addition to facilitating movement. When insects are moving at a slow speed, they move using a leg at a time but keeping the rest in contact with the ground. When insects are moving at an intermediate speed, insects can move their legs two at a time but to maintain stability at least one leg of each segment of its body always remains on the ground.

This movement results in a wave -like pattern known as metachronal gait. An insect can move three legs at the same time when they are running and it is referred to as the tripod gait. The insects have three legs on the ground and that is the front and hind legs of one side of the body and the middle leg of the side that is opposite and it is, therefore, no coincidence that insects possess six legs which are the minimum required for the alternation of the tripod support (Wigglesworth, 2012).

7.2.1 Structure of the Leg

All insects have a total of six legs, this includes pairs of fore, mid and hind legs which are found on the pro, meso and metathorax in that order. Their legs are also divided into six segments. From the top, the segments are called coxa, trochanter, femur, tibia, tarsus and post tarsus. The tarsus has five segments referred to as pseudo since they are connected by a muscle that is similar hence the segments are referred to as tarsomeres. The post tarsus has pulvilli which are smooth or pads that are hairy used in adhesion. The largest segments of the legs are the tibia and the fibula. Depending on an insects way of life, there is a great deal of modification on the insects' legs for example insects that walk and run the cockroach or the stick insects have femurs and tibias that are well developed and in contrast for insects that jump for example, the grasshopper has hind femur and tibia that are disproportionate in size creating the force they require for movement especially for the distance that is long. Swimming insects also possess tibia and tarsi but instead, they bear long hair fringes that function to propel the insects to move through the water. Insects that dig also have a tibia that is modified to dig that is efficient (Shaw, 2014).

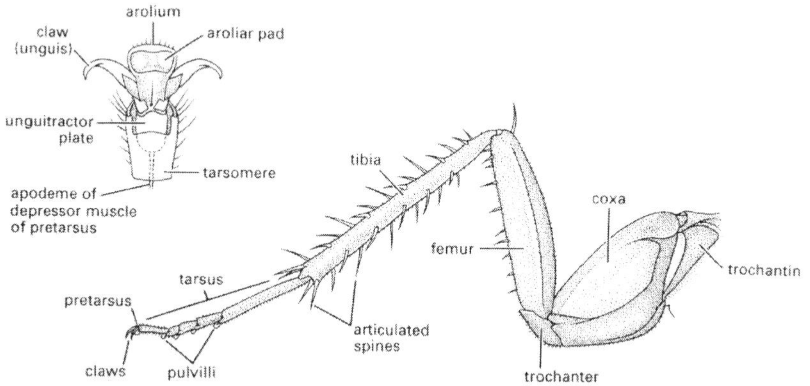

Figure 7.2: Anatomy of the hind leg of a cockroach.

[Source: www.entomologa.ru/outline/22.htm]

*Coxa:*The coxa is the proximal segment and the functional base of the insect leg. The coxa articulates with the pleuron and associated sclerites of its segments on the thorax and in some species the coxa articulates and it similarly articulates with the edge of the sternite. Scientists suggest that the coxa are derived from the sub coxa found on the ancestors of insects. The meron is the posterior lobe which is usually the largest part of the coxa. The meron is well developed in insect species of the Periplaneta, isopteran, Lepidoptera and the Neuroptera.

Trochanter: The coxa also articulates with the trochanter and is rigidly attached to the femur of insects but is movable freely in a direction that is movable on the coxa. The trochanter appearance may be confusing in some insects; in the Odonata species, it has two sub- segments. The base of the femur appears to look like a second trochanter for the Hymenoptera species which are parasitic.

Femur: The femur is the largest part of the leg in most insects. In some insects, the femur is conspicuous especially those with saltatorial legs. This is because the mechanism they use for movement is straightening the joint between their tibia and their femur which contains the massive bipennate musculature which is necessary.

Tibia: In a typical insect leg, the tibia is the fourth section. The tibia is usually generally smaller in comparison to the femur in most insect but then most times it is longer in comparison to the femur. There is a tibial spur near the distal end of the tibia and this can be two or more.in the suborder apocrita the foreleg tibia has apical spurs that are large and fit over a gap that is semicircular in the first segment of the tarsus.it is lined with bristles that are comb- like and insects clean their antennas by drawing them through them.

Tarsus: In the past, the tarsus had a single segment and in the extant, the protura, certain insect larvae and diplura also have tarsus with a single segment and in contrast modern- day insects have tarsus that ate segmented into more than one and can amount up to five. The number of segments will vary depending on the taxon group of the insects and therefore making it easy to identify. An example would be the taxon Pterogeniidae with have forelimbs that are segmented five times and possess a mid-tarsus that is also segmented five times but their hind tarsi are segmented four times. The taxon Cerylodinae on each tarsus has four tarsomeres.

Pretarsus: It is also called the acropod or the post tarsus. The pretarsus is the distal segment of the leg of a typical insect. In some insects, the pretarsus is a single claw such as the collembola, Protura and the larvae of many insects. Most insects have a pair of claws on the pretarsus and these are the unguis and the plural ungues. The pretarsus is supported by the median unguitractor plate which is between the ungues. This plate is attached to the apodeme of the flexor muscle of the ungues. Outside the surface of the unguitractor plate which is between the claws of insects, asymmetrical pair of structures referred to as the parempodia arise in the infraclass Neoptera of insects. It is also present in other classes such as the Hemiptera and also exists in most classes of the Heteroptera. In most species the Parempodia is bristles and in some species, it is also fleshy.in other cases, they are reduced in size so that they can seem to be almost disappearing. The pretarsus also expands forward into the arolium which is the median lobe and is and this happens above the unguitractor plate.

The Embioptera who are web spinners have basal tarsomeres which are enlarged on each of their forelegs and they contain glands that produce silk. For members of the Diptera family, under their pretarsi, they have lobes that are generally paired. Below each unguis, there is a single pair of lobes or pulvilli. Between them, the pulvilli often have an arolium or sometimes a bristole that is in the middle or empodium which means that this is the meeting point of the pulvilli. The plantulae are pulvilli like organs that often exist on the underside of the segments on the tarsus. The arolium, plantulae and the pulvilli are organs that are adhesive and they provide insects possessing them with the ability to climb surfaces that are either steep or smooth. These structures are all outgrowths of the exoskeleton and their cavities contain blood. They are also covered with tubular hair and tenant, the apices of which are moistened with secretions from glands. The organs adapt in such a manner to apply hairs closely to a surface that is smooth so that adhesion occurs through molecular forces on the surface.

7.2.2 Variations in the Functional Anatomy of Insect Legs

The typical thoracic leg of an insect that is in its adult stage is adapted for the functions of running rather than other functions like digging, leaping, swimming, predation and other functions a good example are the legs of cockroaches, none the less there are other functions which legs are adapted to perform which are specialized.

Types of insect legs: Types of insect legs can either be ambulatory legs saltatorial for jumping, raptorial for seizing prey, fossorial for digging, natatorial for swimming and cursorial for running, skating legs, perching legs, clinging and carbiculate legs.

- Ambulatory legs-the legs are slender and elongated and are designed to walk. The insect legs are clawed at the end and have cushioned pads that allow them to stick on smooth surfaces they include insects such as the housefly.
- Cursorial legs- they are legs modified to run. The leg segments are long and thin.

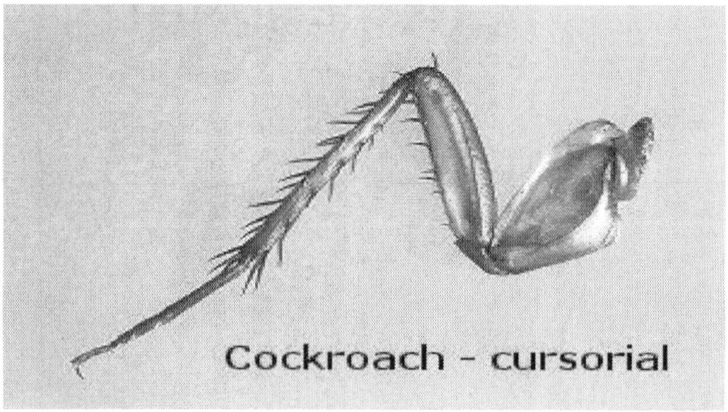

Figure 7.3: The Cursorial leg of a cockroach.

[Source: https://genent.cals.ncsu.edu/bug-bytes/thorax/legs/]

- Saltatorial legs-the has large hind limbs that are adapted for jumping. The femur and the tibia are both elongated.

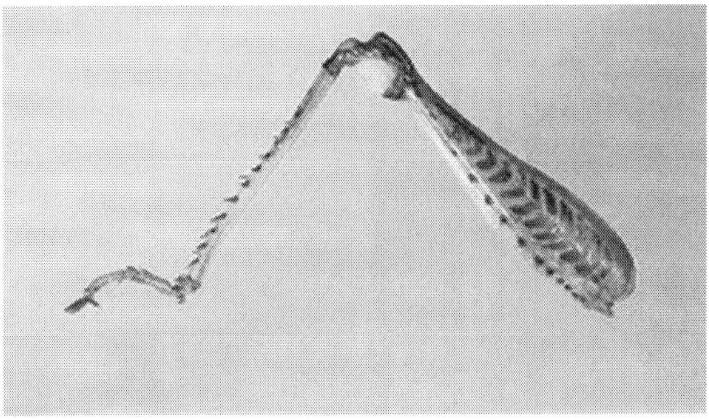

Figure 7.4: The saltatorial leg of a grasshopper.

[Source: https://genent.cals.ncsu.edu/bug-bytes/thorax/legs/]

- Fossorial legs- insects with the legs have forelegs that are spade like and they used for burrowing and digging rapidly.

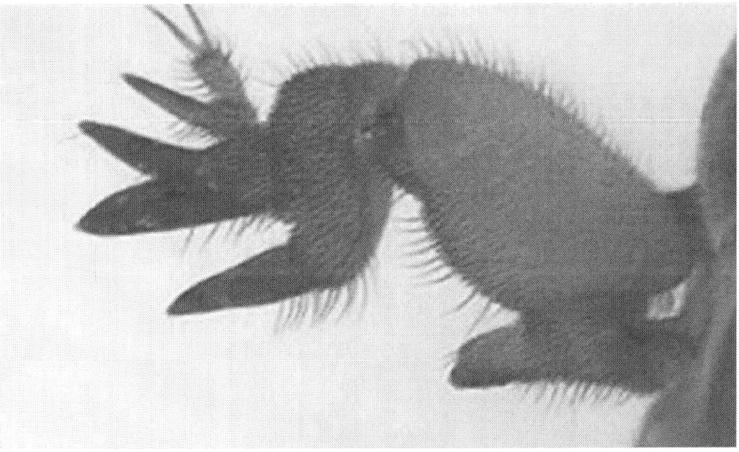

Figure 7.5: The Fossorial leg of the mole cricket.

[Source: https://genent.cals.ncsu.edu/bug-bytes/thorax/legs/]

- Natatorial legs-this legs are flattened and fringed and are found in insects that are aquatic for swimming purposes. These insects have setae that are long on the tarsi that function as oars to paddle through the water.

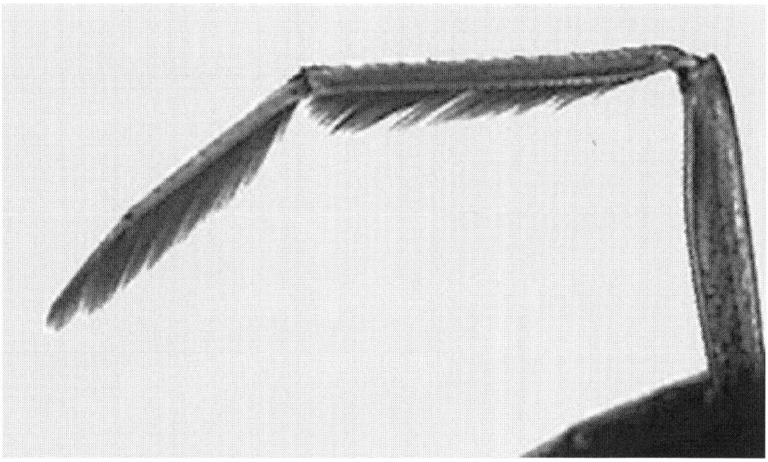

Figure 76: The Natatorial leg of the backswimmer.

[Source: https://genent.cals.ncsu.edu/bug-bytes/thorax/legs/]

- Skating legs- are long legs that also have hydrophobic hairs and claws that are ante apical to skate on water surfaces.
- Perching legs-they have legs that are spiny for piercing, perching and seizing and also holding their prey and these legs are also not strong enough for walking
- Raptorial legs-this insect legs are long and stout and are designed for grasping.

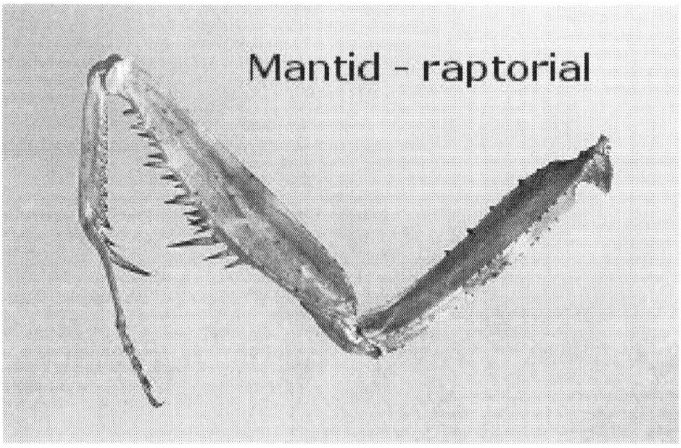

Figure 7.7: A mantid's raptorial leg for seizing prey.

[Source: https://genent.cals.ncsu.edu/bug-bytes/thorax/legs/]

- Clinging legs-on the tip of each leg there are claws that are sharp and this improves clinging to the hair of the host.
- Carbiculate legs-there is three structures on the hind legs of insects that are pollen baskets that have hairs, pincers that are heavy with wax and a pollen combat the tip.

Functions of insect legs

- The legs play an important role in the defense of insects not only by allowing them to run, jump, burrow and swim but it also allows them to kick or slash their prey.
- Legs also have spines that they use in defense and they deter their predators or competitors and can cause considerable damage.
- Stink bugs and treetops are insects that deliver kicks that are powerful at predators who attempt to attack their young ones.
- Legs can also have sensory structures that allow them to feel, hear or taste which allows insects to assess their environment initially.
- Insects can also use their legs to in the production of sounds for communication that is intraspecific by drumming their legs against their substrate.
- Insects also use their leg to preen and groom to maintain their hygiene. They use various structures in their legs such as combs that are cuticular, setal brushes, grooves and notches.
- Honey bees also have hind legs that are also specialized to groom pollen from the plumose hairs of their body.
- Legs are also used for grasping purposes especially for insects that have legs that are raptorial and it is seen especially in insects that are predators
- At the base of legs for insects such as houseflies, they have lobes that are thumb like and also claws that are talon- like that help them grasp on to their host in the event they are feeding off of them.
- Leg patterns and configurations play a role that is important in camouflaging, mimicry and also courtship.

Other Examples of leg specialization in insects include;

- Burrowing- the front legs of the Gryllotalpidae and some members of the superfamily Scarabaidae are adapted for earth burrowing purposes.

- Seizing and holding pray- this applies to the forelegs of the Mantodea, Phymatinae and Mantispidae being adapted for the function of seizing and holding their prey in one way and in contrast members from Gyrinidae are adapted to holding or catching their prey in a different way.
- Walking-in some butterflies, the forelegs are reduced for example the Nymphalidae that only two out of the six remain for the function of walking.
- Leaping-some insects possess hind legs that are saltatorial. These insects possess femurs that have bipennate muscles and are straight and a tibia that is long for purposes of leaping and sometimes kicking or pushing as a form of defense. Members of the subfamily Halticinae such as flea beetles have a hind femur that is powerful and allows them to leap in a spectacular manner. Other beetles may also have spectacular leaping but may not have hind limbs that are saltatorial but on the other hand, can be clumsy instead for example, some members of bruchinae use their hind limb that is swollen to push their way out of plants seeds with shells that are hard such as the erythrina in which they grew in into adulthood.
- Insect species such as the Odonata, dragonflies and damselflies have their legs adapted for the function of seizing their prey which they feed on in the process of flying or even while sited still on a plant. The insects are almost unable to use their legs for walking.
- Insects that inhabit water most of the time use their legs for swimming and some immature insects use other means for movement such as wriggling, undulating and even expelling jets of water.
- Crane flies have basitarsus which are inflated and contain tracheal sacs to increase the buoyancy of insects that facilitate drifting while flying.
- Some insects also use their legs like a second set of antennae and this includes members of the Chironomidae and similarly, these are seen in groups that completely lack antennas such as the Protura.
- The basitarus of some insect legs also house silk glands and each of these glands is connected to a seta with a core that is apical through which the silk is also extruded.

Muscles of the legs

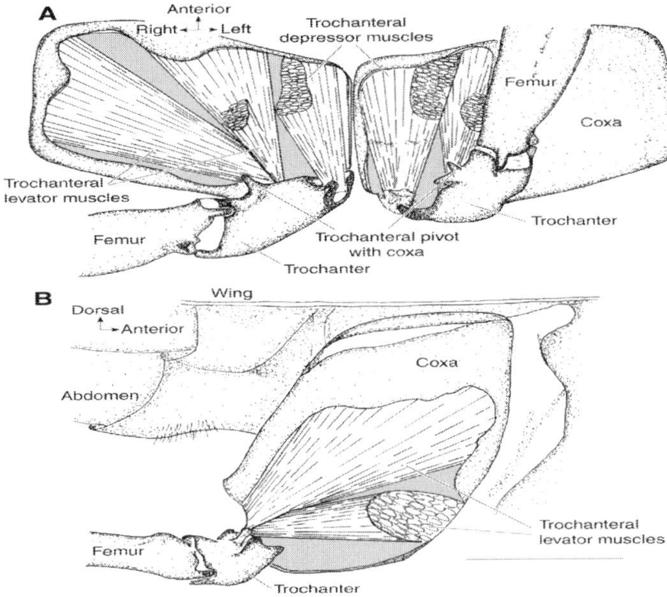

Figure 7.8: Anatomy of the moving muscles of the hind leg coxo-trochanteral joint of *Aphrodes*. A) The hind leg fully elevated in a view that is ventral and a hind leg on the right depressed partly. The trochanteral depressor muscle of each hind leg is medial and the different parts of the levator muscle are more lateral. (B) The right metathorax view on the side exposed to show the trochanteral levator muscles that are placed laterally on the right hind leg in a position that is depressed.

[Source: https://jeb.biologists.org/content/210/20/3590.figures-only]

Muscles that move legs fall into two categories and they are either intrinsic or extrinsic. The coxa of the legs is moved by muscles that are extrinsic and they arise from the thorax. They include the promotor and motor muscles that arise from the tergum and also the abductor and adductor muscles which originate from the pleuron and sternum; finally, rotator musclesthat originate from the sternum. The role of the muscles will vary depending on the type of movement and also the articulation. For example, in Apis, which is a member of Hymenoptera which have plural or sternal applications that are rigid, there is an absence of the promotor and the remotor muscles from the tergum. The intrinsic musculature is simpler than that of the coxal musculature. It typically consists of pairs of muscles that are antagonistic in each segment. In species of the Blattodea, the trochanter has three elevator muscles that arise from the coxa and three depressor muscles, two also that originate from the coxa and a third which arises from the pleural ridge and the tergum.

The trochanter is immovably attached to the femur. The tibia is moved by extensor and flexor muscles which arise from the femur. In the tibia, there are levator and depressor muscles arise from the tibia and are attached to the proximal end of the basitarsus but tarsus moving to the tarsomeres lack muscles. For most insects, the pretarsus has depressor muscles but lack levator muscles. The fibers of depressor muscles occur in groups that are small in the femur and tibia and they are attached to into an apodeme that is long and arises on the unguitractor. Levation of the pretarsus leads to the elasticity of parts that are basal (Herreid, 2012).

They are complexities in the innervation of muscles in the legs. Most muscles are innervated by axons that are either fast or slow and also inhibitory but not also fibers found within the muscles are innervated by the different types of motor neurons. For example, for members of the Peripleneta, each of the meso and meta thoracic coxae have four muscles which depress the trochanter. Two of these muscles are innervated by the fast axon only and it also supplies the other two muscle parts.

The parts are also innervated by a slow axon which also supplies to other parts that are on these muscles that lack fast nerve supply. Those parts of the muscles that receive input from the slow axon and not the fast axon are innervated by three inhibitory fibers. There is a more complex supply in the extensor tibia muscle of the hind leg of a grasshopper.in addition to the fast, slow and inhibitory axons, it receives input from the axon which releases neuromodulator octopamine. This axon is known as referred to as the dorsal unpaired median axon of the extensor tibia muscle. The fast axon innervates a majority of fibers and may not be generally a characteristic of muscles in the leg, as the muscles of the extensor tibia of grasshoppers are specialized for jumping. A common inhibitory neuron innervates most of the muscles of a coxa that is moving, the trochanter and the tibia that is there in each leg of locusts or grasshoppers. Two other inhibitory neurons innervate the muscles of the distal parts of legs, which both run to similar muscles (Herreid, 2012).

Sensory system of legs: Insect legs have sensory systems that are extensive. Some of the sensory elements include proprioceptors which monitor the positions of the leg segments and the insect stance.in the perception of environmental stimuli, other mechanoreceptors and chemoreceptors are involved. Proprioceptors found on the leg include hair plates, campaniform sensilla and organs that are chordotonal. The gravitational sense of the leg is contributed to by hair plates. Periplaneta has hair plates at the proximal end of the coxa and also the joint of the coxa trochanter. On the trochanter, there are groups of the campaniform sensilla, there is also a group on the femur proximally and another on the tibia; a small number also on the dorsal surface at the distal end of each tarsomere. For each front leg, there are about 140 sensilla in hair plates 80 campaniform sensilla in total.

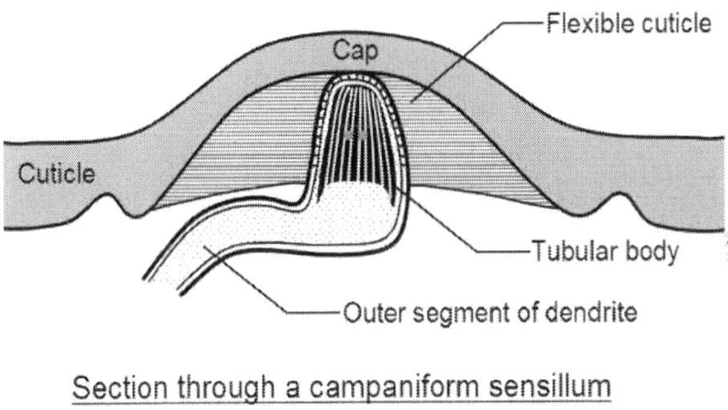

Figure 7.9: A cross- sectional view of the campaniform sensillum.

[Source: https://cronodon.com/BioTech/insect_mechanoreceptors.html]

In addition to this, they have an organ that is chordotonal associated with each joint leg and also have multipolar neurons that at the trochanter-femur and femur-tibia joints. There may also be strand receptors. The same arrangements are in locusts that migrate in the case of proprioceptors. This can also be seen in locusts, the stick insect, carasius and the adult tobacco hornworm. On the legs of insects, many sensilla are exteroceptive. Insect legs also have mechanosensitive trichoid sensilla that are distributed all over. On each tarsus, the final larvae stage of the orthoptera has about 140 of sensilla on each tarsus and similarly, others are present on the femur and tibia. Within the central nervous system, the axons from the sensilla in different regions on the leg converge to interneurons that are separate so that information that is spatial for maintenance. This is only true for the first order spiking interneurons that are local but also interneurons that are spiking and intersegment neurons that are spiking. This ensures that the insect responds in a manner that is inappropriate when a particular part of the leg is touched (Herreid, 2012).

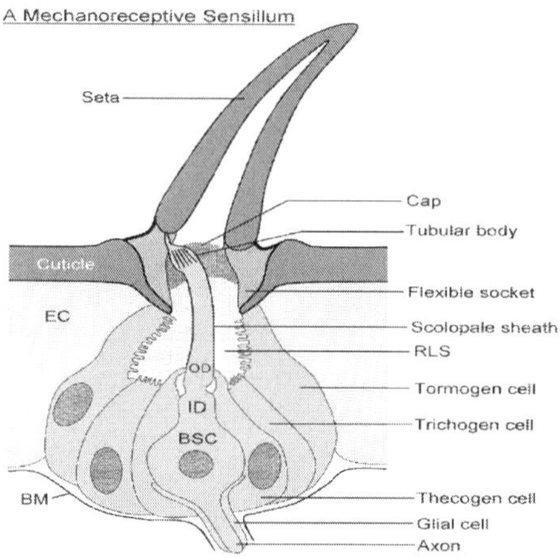

Figure 7.10: A diagram representing a mechanoreceptive sensillum.

[Source: https://cronodon.com/BioTech/insect_mechanoreceptors.html]

Each leg also has multiple contact chemoreceptors. *S. americana* has about 200 chemoreceptors on the upper surface of the tarsus and over 100 chemoreceptors on the pulvillar pads on which the insect normally stands. Despite many chemoreceptors being present on the tarsus, they can also be present on the femur and the tibia. Holometabolous and hemimetabolous insects in their adult stage have tarsal chemoreceptors but there is a lack of evidence that they exist on the legs of holometabolous insects. Mechanosensitive hairs exist on the legs of caterpillars in small numbers. Most insects also have an organ that is subgenual and is also sensitive to substrate vibration.

7.3 MECHANISMS OF LOCOMOTION IN INSECTS

7.3.1 Maintenance of Stance

Muscular activity is also required when standing and adjustment that is continuous of this activity that compensates when there is a shift in the position of the insect. Muscles can be continuously active even in an insect that is stationary for example the extensor muscles of the tibia of the fore and mid limbs of the schistocera. Axons innervating slow muscles can fire at low rates, with their activity varying with the position of the leg. In a stationary insect, the fast axons are not active and the inhibitory neurons are only active sporadically but do fire in response to any contact or vibration. The postural position reflex is activated when an insect standing still experiences a change in angles as a result of a force created by an example of a sudden gust of wind. Variations in the functional anatomy of insect legs. This is mediated by organs that are chordotonal (Dudley, 2002).

Studies were done to examine the joint between the tibia and fibula show that in the case there is the tendency to reduce the angle between them, there is opposition by the extensor muscles found on the tibia while an increase is opposed by the tibial flexor muscle. Similar reflexes are also described at other joints in the leg. The reflexes are produced in response to changes that are passive at angles that are between segments on the legs, whatever the initial degree of the extension or flexion. The input from the organ that is chordotonal and the activity of the motor neuron to the muscles varies depending on the position. When locusts are standing on a horizontal surface, for the forelegs the angle between the tibia and the femur is usually close to ninety degrees and for the hind legs the angle varies but it is usually less than forty-five degrees. Reflexes related to postural resistance contribute to the maintenance of these positions in the mid and hind legs but not the front legs. The campaniform sensilla on the legs of the insect monitor strain in the cuticle and produce reflex responses in muscles tending to alleviate those strains (Dudley, 2002).

Some campaniform sensilla at the proximal end of the tibia of Periplaneta are oriented parallel with the long axis of the leg; others at right angles to it. The sensilla sense axial forces that would have been produced by the weight of insects that are oriented proximally or transversely although these forces are also observed to be relatively weak. The tibia bending produces strong responses. When the insect is standing on a surface that is horizontal with the tibia inclined away from the body, the mass of the body tends to cause bending upwards of the tibia, compressing the proximal sensilla. Both groups of sensilla respond when there is twisting of the tibia. Stimulation of the proximal sensilla excites slow motor neurons to the tibia extensor and the extensor muscles of the trochanter and this inhibits the motor neurons that are slow to the flexor of the tibia and the flexor muscles of the trochanter. An opposite effect is observed when the distal sensilla are stimulated.

A reflex system that is comparable and compensates for stress is in the trochanter of the stick insect when the femur is bent both anteriorly and posteriorly. In this case, there are two groups of the campaniform sensilla on the trochanter which activate the retractor and protractor muscles of the coxa reflexively. The pathways between proprioceptive sense cells and motor neurons are in some cases at least monosynaptic. Similar systems for stance control is seen in most insects (Herreid, 2012).

7.3.2 Patterns of Leg Movement

Dividing steps during walking or running are divided into two phases. During the phase of stance, the leg is retracted such that it moves backward relative to the body with its foot on the ground, therefore, allowing the insect to move forward. During the swing phase, the leg is protracted so that it swings forwards with the foot off the ground. The stepping pattern often varies with the speed of the movement. When insects are moving at a slow speed, it has most of its feet on the ground for a lot of time. In this case, the legs may be singly protracted in a wave that is metachronal and runs first down one side of the body and then similarly down the other in the direction that is anteroposterior.

At stepping rates that are higher, insects adopt a tripod gait that is alternating in which the body of the insect is supported always on a tripod formed by the fore and hind legs of the other. As one set of the tripod gait is protracted the other is retracted. At low speeds, both sets of legs may also be planted on the ground and at speeds that are high both sets of legs can be off the ground at the same time. At very high speeds insects such as the cockroaches can alternate between a gait that is quadrupedal and bipedal. This means that they run on their hind legs and raise their forelegs at an increasing angle to the ground.

Some insects are quadrupedal due to modifications of both the fore and the hind legs for other purposes for example mantids most of the time walk with their front legs on the ground and this is also common for grasshoppers in which the hind leg is modified for jumping and uses only the anterior parts of the legs for walking. At high speeds, Orthoptera move their legs of each segment in sync and also maintains its stability by using the abdomen tip as an additional support point.

Increasing the frequency of leg movement leads to the achievement of higher walking and running speeds. Stride frequency is mainly increased by shortening the duration of the phase of stance.

7.3.3 Coordination of Movement in Insects

The coordination of leg movements in insects is regulated by a neuron network that produces an output that is rhythmic without the need for any external timing signals. Such networks are similarly referred to as central pattern generators. There is at least a central pattern generator for each leg. Interneurons link the individual networks together and the output from each central pattern generator is modified as required by the sensory feedback from the legs. Only animals that possess a body frame that is rigid use the tripod gait movement. Insects that have soft bodies such as caterpillars have a hydrostatic skeleton. Their movement is by contractions that are peristaltic of the body, pulling the hind legs forward to grab the substrate and then push their front body forward segment to segment. The movement is exaggerated in the larvae of geometrid moths. In the process of grabbing their substrate with their thoracic legs, they hunch the abdomen towards the thorax, grasping the substrate with their prolegs and then extending their anterior end as far as possible. It has adopted nicknames such as inchworms, spinworms or measuring worms.

Different segments of the leg have antagonistic muscles and their activity leads to alternating movements of protraction and retraction. In some species for example, Periplaneta, the trochanters levator muscles are active during protraction hence lifting the foot of the ground contrary to this the depressor muscle is inactive but it starts to contract before the insect places its leg on the ground while the lavatory muscle is still active. When the foot is on the ground the contralateral depressor muscle remains active when the foot is on the ground. This has similarly been reported in other insects.

The speed of movement determines the innervation of the muscles by the fast and slow motor neurons. When an insect is moving at a slow pace, the angular velocity of the joints in the phase of stance will vary and this is in proportion to the average fixing rates of the slow motor neurons that innervate depressor muscles of the coxa and the extensor muscles of the tibia. Controlling the running speed and direction by the modulation of the average firing rate of slow motor neurons. Variation of the firing rate within a burst does not appear to impact upon the joint movement.

In the Blaberus insect species, both slow and fast motor neurons are utilized but this does not result in any acceleration that is sudden by the leg. The angular joint velocity will vary continuously depending on the average firing velocity of motor neurons that are slow during the phase of stance. The firing of the fast coxal depressor motor neuron is timed to coincide with the transition from flexion to extension and it is associated with a transition between the two phases of the step cycle.

Studies have been done on how the patterns of movement are generated and coordinated. In locusts, the metathoracic ganglion which is isolated is capable of generating activity that is rhythmic and alternating in pools of antagonistic motor neurons when treated with muscarinic agonist pilocarpine similarly in Periplaneta the central nervous system is capable of producing bursts that are rhythmic for an activity that is reciprocal in the levator and depressor motor neurons of the coxa even when there is an absence of sensory input on the leg (Herreid, 2012).

7.3.4 Interleg Coordination

To generate an effective tripod gait, insects must coordinate joint movements not only with a single leg but also between legs. On the surface, inter-leg coordination appears to be a simple job. To make a tripod gait by the three legs, the tarsus must move in a manner that is synchronized. The legs of each segment make movements that are different and so do the patterns thus the insect must coordinate the tarsi through joint movements and neutral activities that are specific to each pair of legs. Observations have been made in relation to the stick insects and it suggests that events leading to the movement of one leg and a readily detected by leg sensors influence the actions of other legs in ways that are discrete for example if an insect swings one leg, the other anterior leg that is adjacent cannot also enter into a swing phase. It can also be said that the beginning of a stance phase in one leg triggers a swing phase in the anterior adjacent leg and in the leg that is the contralateral leg of a similar segment. More influences that are subtle are also present to account for changes that are transient for example, an increase in the load on one leg prolongs the power stroke on the contralateral leg from the same segment. This results in the co-activation of legs when an insect encounters an increase in resistance which can be experienced when it is climbing (Herreid, 2012).

7.3.5 Negotiating Obstacles

The remarkable agility that attracts robotic interests by scientists is seen when insects meet an obstacle. Most insects can negotiate obstacles in a manner that is effortless some of which would cause considerable problems if it was in the case of controlling robotics. Climbing over obstacles is considered transient in comparison to when the insects are moving on a ground that is horizontal and the strategies insects use will, therefore, vary depending on the nature of the obstacle however some themes are recurring for example if insects meet an obstacle for example, a block, they tend to move their Centre of mass upward to surmount their obstacle. For example, in the case of cockroaches, the strategy it will apply will depend on the obstacle size and the speed of the moving insect when the block is encountered. The forelegs are normally lifted fairly high during the time they are walking normally. If the insect encounters an obstacle the lower than the normal front leg trajectory, then the insects do not need to alter their movement at all. The front legs will be placed on the obstacle and moved down to push the insect's Centre of gravity upwards as a natural consequence of encountering an obstacle (Noest, 2017).

If an insect encounters blocks that are large, it will require redirection of leg movements that push the insect body upwards upon extension of the joint. This allows the forelegs to be placed rapidly on top of the block to allow commencement of climbing. For very large obstacles, the insects must attach themselves. The problem is solved by combining claws and pads that allow adherence to the substrate. The claws of cockroaches are held up during movement on a surface that is horizontal. When it walks on a vertical surface, the claws, in contrast, are pulled down by tibia muscles, which attach to claws through a long apodeme.

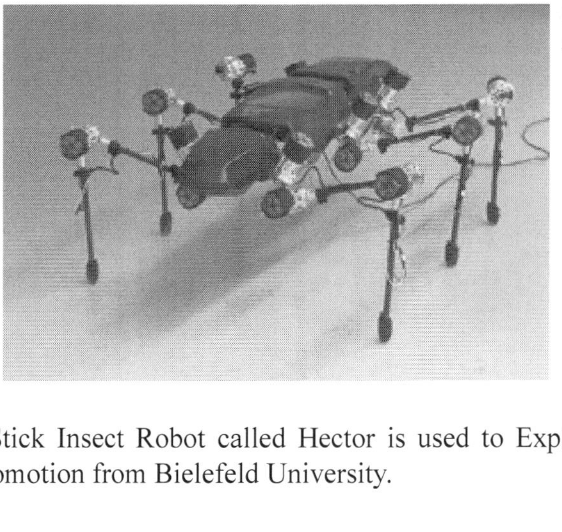

Figure 7.11: Stick Insect Robot called Hector is used to Explores different models of Locomotion from Bielefeld University.

[Source: https://www.botmag.com/stick-insect-robot-explores-different-model-for-locomotion/]

7.4 CRAWLING IN INSECTS

Larvae of insects that are holometabolous move by changing their shape rather than their leg movements as in adult insects that walk or run. This type of locomotion can be referred to as crawling. In a majority of the forms of crawling, the cuticle is soft and flexible and therefore does not by itself provide a skeleton that is suitable for and on which muscles can act. Instead, the body which contains hemolymph provides a skeleton that is hydrostatic. Turgidity on the walls of the caterpillar is maintained by muscles and because of the incompressibility of the fluids in their body, compression of one part of the body due to contraction of the muscles is compensated for by the expansion of other body parts. The place and form of these compensating changes are controlled by the differences in tension of the muscles throughout the body. Lepidoptera caterpillars have a pair of prolegs in addition to the thoracic legs that are on the abdominal segments three to six and another pair on its tenth segment. The prolegs are outgrowths from the walls of the caterpillar. The apical area is less rigid on the sides also known as the plant and it bears one or more rows of circles of hooks that are curved outwardly or crotchets with which the prolegs obtain a grip (Price, 2003).

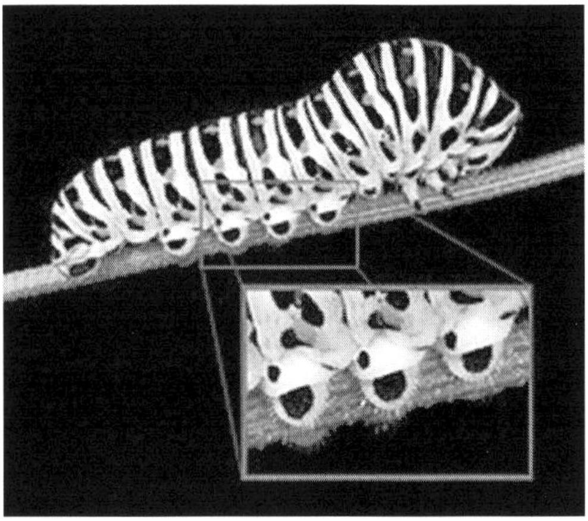

Figure 7.12: Lepidoptera: *Papilio machaon* caterpillar with 5 pairs of prolegs.

[Source: https://en.wikipedia.org/wiki/Proleg]

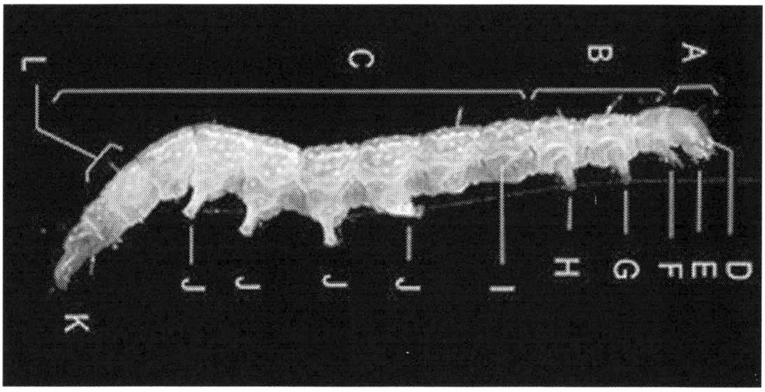

Figure 7.13: Medial prolegs labeled *J*, anal proleg *K*. (The true legs are *F*, *G* and *H*.).

[Source: https://en.wikipedia.org/wiki/Proleg]

The Centre of the planta has retractor muscle insertion such that when they contract it is drawn inwards and the crotchets are disengaged. The leg is evaginated by turgor pressure when the muscles relax. Prolegs can function as suckers when they are on smooth surfaces. The crotchet is turned up and the surface of the planta is pressed first down onto the substratum and then the center is drawn slightly up so as to create a vacuum. The movement of caterpillars occurs by serial contractions of the muscles that are longitudinally coupled with leg movements that begin posteriorly and this continues as a wave to the front of the body. Two legs of each segment include the thoracic segment move in sync. The dorsal longitudinal contraction lifts each segment and at the same time, the prolegs are retracted (Price, 2003).

7.5 JUMPING/HOPPING IN INSECTS

Insects can jump vertically upwards at 45 cm and 90 cm horizontally. This is equivalent to a human jumping 20 meters and jumping long at 40 meters. Insects can also carry objects of up to two or three times their weight and for ants up to between seven to ten times. The strength of leg muscles increases with their cross- sectional area and their thickness.

The force created by a grasshopper jumping is about four times in comparison to that generated by a human leg. The cuticle of the insect contains elastic material called resilin which is abundant in insects that jump and it enables the legs of insects to store energy. In the insect's preparation to jump, the leg is locked in place by powerful flexor muscles of the tibia which contract, straining the cuticle of the leg that is locked which builds elastic energy. When the cuticular lock is released, the leg springs straight releasing energy that is stored and elastic as the insect jumps. Ninety-seven percent of the stored energy is mechanical and the remaining is lost as the energy that is unusable (Dudley, 2002).

Issus is a genus of plant hoppers known to possess the gear mechanism which is used in locomotion. They are insects that are generally flightless. The first form of the mechanism was observed in the *Issus coleopterous* species. The gears keep the legs in sync and allow insects to jump up to 400 gs in two milliseconds. The existent of gears has been known for decades but until recently characterization of their functional significance has been done. The gears are observed in nymphs of plant hoppers and it is suspected that the mechanism is lost in adults after their last molt because if it is broken it cannot be repaired thus crippling the insect for life. Before this discovery was made it was assumed that gears were created and used by humans (Dudley, 2002).

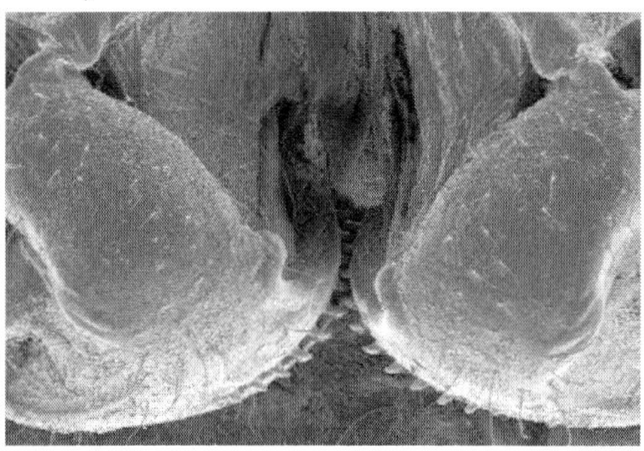

Figure 7.14: The gear system in plant hopper insects.

[Source: home.bt.com/news/odd-news/gear-system-found-in-insect-legs-11363833068993]

7.6 FLIGHT IN INSECTS

Insects are the only invertebrate group that possesses wings and flight that have evolved. The flight of insects began in the Carboniferous period about three hundred and fifty years ago. Only two groups of insects that fly have flight muscles directly attached to their wings that is the dragonfly and the mayfly. In other insects with wings, the muscles used for flight are attached to their thorax. Development of flight in insects has allowed them to have better mobility and this has improved an insect's ability to access food, mate and also improved dispersal powers. Having the power to fly has also allowed insects to exploit new environments. Accessibility to plants with flowers has been made possible in insects that possess wings. In adult insects' wings are fully functional and are also visible in developing nymphs. Two pairs of wings are functional and arise dorsolaterally. This is as fore wings on the second and on the third segment of the thorax as the hind wings (Dudley, 2002).

7.6.1 The Insect Wing

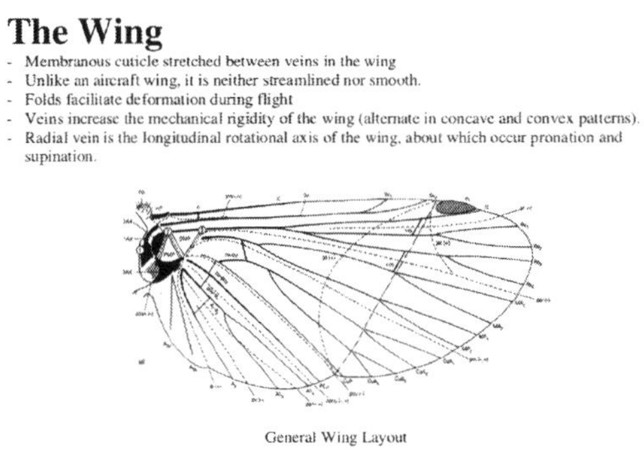

Figure 7.15: The general layout of the wing.

[Source: https://www.slideshare.net/MItalo/flight-of-insects.]

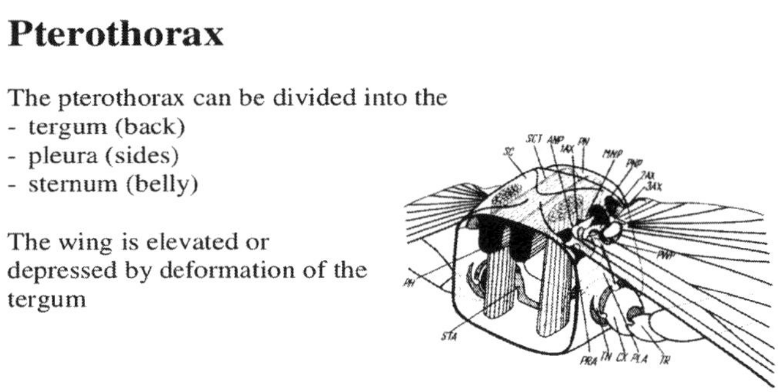

Figure 7.16: Anatomy of the pterothorax.

[Source: https://www.slideshare.net/MItalo/flight-of-insects]

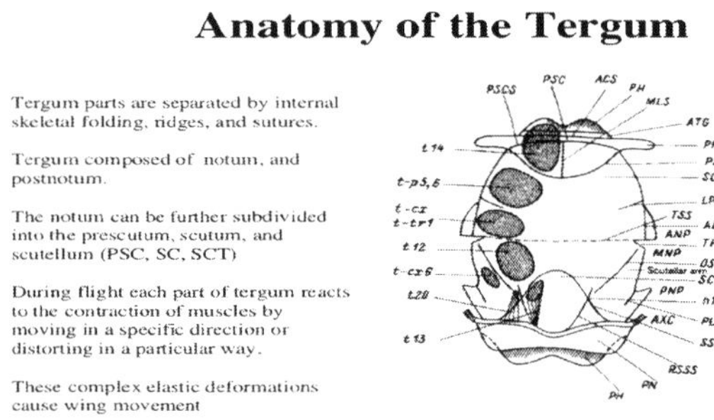

Figure 7.17: Anatomy of the tergum.

[Source: https://www.slideshare.net/MItalo/flight-of-insects]

For insects to fly, they must overcome the forces of gravity and air resistance. In gliding flight where the wings are outstretched rigidly, insects overcome the forces through the use of air movements that are passive known as relative wind. Insects are lifted by adjusting the leading edge of the wing when it is oriented into the wind. As the angle increases, so does the lift until stalling occurs

Figure 7.18: A moth feeding mid-flight.

[Source: https://en.wikipedia.org/wiki/Insect]

Most insects glide a little while others glide extensively such as dragonflies and some species of the grasshopper. Most insects fly by beating their wings and examining this may be difficult. Since the frequency of a large slow- flying butterfly maybe five times per second a bee maybe 180 Hz. A single beat of insect wings comprises three interlinked movements.

The first cycle is downward, a motion that is forward and then upward. During the cycle, the wing is rotated around its base. The third component occurs during flexing of various components of the wings and this occurs in relation to the variation in air pressure. In true flight, moving winds produce relative wind and in contrast in contrast to gliding. An insect that is flying makes adjustments that are constant so that air which is in front of the insect moves downwards and backward during a wing beat which causes the insect to move upwards and forward (Dudley, 2002).

Mechanisms for insect flight are not complicated. Wing bearing segments on the thorax can be envisaged as a box with sides the pleura and a base called the sternum and it fused rigidly. Wings connect where the rigid tergum is attached to the pleura by membranes that are flexible. This attachment which is made up of a membrane contains resilin which provides the thoracic box with elasticity. Insects that fly have two different kinds of muscle arrangement which are used to power flight. These are the:

- Direct flight muscles- which is connected to the wings
- Indirect system- where there is a lack of muscle to wing connection but rather the action of muscles deform the thoracic box for wing movement.

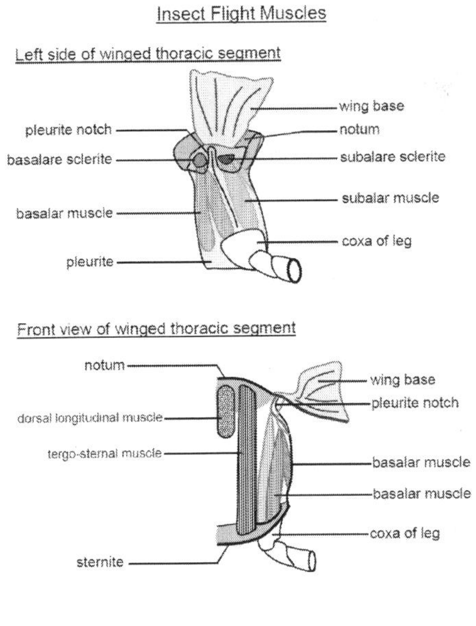

Figure 7.19: Insect flight muscles.

[Source: https://www.cronodon.com/images/insect_flight_muscles_large.jpg]

Odonata and Blattodea, which are old insect groups appear to use direct flight muscles to degrees that vary although at least some recovery muscles may be direct. Modern insects use muscles that are indirect for flight and provide wing orientation rather than the production of power. Muscles attached to the wing base contract to produce an upward stroke on the point that is pivotal on the other hand the downward wind stroke is produced through muscle contraction that extends from the sternum to the wing base and this is outside the point of pivot (Dudley, 2002).

Indirect flight muscles are attached to the sternum and tergum. Contraction causes the tergum and the base of the wing to be pulled down. This movement levers the outer and main part of the wing to move in an upward stroke. The second set of muscle power the downbeat and run from front to back of the thorax consequently deforming the box and forcing the tergum to lift. In all stages of flight, there is relaxation of the flight muscles thus energy is conserved because thorax elasticity restores its shape.

Independent control of the wings can be control by variation of time and rate in the flight direction. Variation in excess may impede flight that is controlled. For insects that have slow wing beat frequency maintenance can be done by muscles that are synchronous and fast- beating muscles require asynchronous muscles. In such insects, the wing is such that the wing has only two stable positions fully up and down. As the wing from one extreme to one which is alternate it passes through an unstable position that is intermediate. At the point of instability elasticity of the thorax snaps the wing through to an alternate position that is stable (Dudley, 2002).

Insects with the asynchronous mechanism have fibrillar flight muscles that are peculiar and possess ta property such that on sudden release of tension by the muscle, the contraction that is next is induced. Oscillation of muscles can also occur and contract at a frequency that is high than the nerve impulses which only need to be periodic to maintain flight by the insect. The rigidity of the thorax also maintains wing harmonization when the tergum becomes depressed or relaxes and this happens in both wings. Retaining of direct flight muscles happens in insects with indirect flight muscles that are used in making adjustments that are fine in the orientation of wings during flight.

Insects use their eyes and their antennas to sense direction and deviations that result from movements in air. Some insects such as Diptera have sensory equipment that is sophisticated as their hind wings are modified for the function of balancing. Flight initiation involves legs springing the insect into the air. The neural firing of the direct flight muscles is caused by loss of contact of tarsals with the ground. Flight activity originates in the contraction of the mid -leg muscle in insects. All of these propel the leg downwards and simultaneously pulls the tergum downwards to inaugurate flight. Legs are also important in the landing process because there is no braking that is gradual by forwarding movement. All the shock is taken from the legs that are outstretched and possess pads, spines and claws for purposes of adhesion (Noest, 2017).

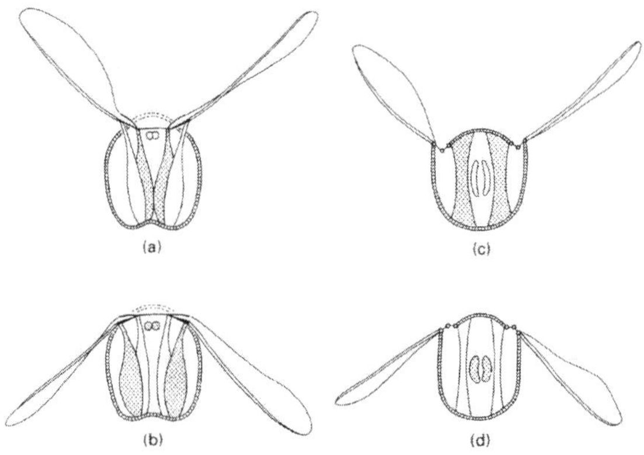

Figures 7.20: Direct flight mechanisms: thorax during (a) upstroke and (b) downstroke of the wings. Indirect flight mechanisms: thorax during (c) upstroke and (d) downstroke of the wings. Stippled muscles are those contracting in each illustration. (After Blaney 1976).

[Source:www.entomologa.ru/outline/32.htm]

7.7 MECHANISMS AND FUNCTIONS OF LEGS IN AQUATIC MOVEMENT

A large number of insects spend part of their life living in water. In some insects that are primitive, their immature stages are spent in an environment that is aquatic and some spend their adult life in waster for example certain water beetles. Most of the insects that move on water belong to the family Gerridae. These species have adaptations for locomotion underwater. Insects such as water beetles and bugs have legs with pad- like structures. Jet propulsion is done by dragonfly naiads by expelling water using force out of their rectal chambers. Some insect species have the capability of walking on water such as water striders. They do this by using their claws which are not at the tip of their legs like other insects but are recessed in a groove that is special and it is there further up their legs. This modification prevents them from piercing the film on the surface of the water the rove beetle Stenus emits secretions from the pygidial gland bringing forth the possibility for them to move on water surfaces by propulsion (Herreid, 2012).

Figure 7.21: The backswimmer *Notonecta glauca* underwater, showing the adaptation of having hind legs that look like paddles.

[Source: https://en.wikipedia.org/wiki/Insect]

Insects use surface tension to their advantage allowing them to float on water. The feet of insects facilitate the creation of indentations in water surfaces and this then allows them to spring back propelling the insect forward. Water striders have bodies that are hairy create buoyancy by trapping air between them and the hairs also repel water and this combined allows the insect to float. They also have legs that are long especially the mid and hind legs in comparison to the forelegs and they use them for propelling over the surface of the water (Herreid, 2012).

Figure 7.22: An insect floating on water as a result of surface tension.

[Source: https://insectcop.net/how-insects-walk-on-water/]

8

Insect Ecology

CONTENTS

8.1 INTRODUCTION

Insect ecology is defined as the scientific study of the existing relationship between insects and their interaction with the environment and their surroundings, also known as the ecosystem (Schowalter, 2006). Ecology is considered to be integrative and this study requires Professional contributions from geologists, Geographers, Mathematicians, Soil scientists, biologists as well as chemists and climatologists all working together to be able to understand the environment affects organisms, population and communities.

An ecosystem is defined as a community formed by living organisms and nonliving components of the environment creating an interactive system containing biotic and abiotic components that are linked together by the existing nutrient cycle and flows of energy. Insects are considered to play a very significant role in the functioning and the structure of the world's environment due to their Form diversity, their lifestyle and functioning, significant considerable biomass, as well as their interaction with organisms in the environment and plant life. Because of their major contribution to the existing biodiversity in a lot of habitats excluding the sea, they equally play a very important role in the ecological system due to their many functions in an ecosystem. In the ecosystem, they represent very significant food resources. They also represent disease vectors for a lot of organisms such as human beings and animals and in addition to that they are very capable of altering rates and directions that energy takes, and these alterations pause serious threats and could lead to the eventual affecting of global processes. Some ecosystems experience I representation of dominant pathways of energy and matter flow by the insects and other arthropods as well as the risks that biomass of these insects could increase beyond control leaving consumption of vegetation that is important to life sustainability and the environment's normal functioning ash risk (Schowalter, 2006).

A variety of insect species have the capability of removing entire vegetation in different sites the minute they are present there. insects affect the environment and more often than not also affected by environmental issues such as diverse ecosystem health, the quality of air and water present, crops that are genetically modified, the Epidemiology of diseases, wildfires, other environmental disturbances, the introduction of exotic species and also their control in an ecosystem, land use and climate change. The study of insect ecology also brings out the important role that is played by insects in maintaining community structure and composition when it comes to transmission of animal diseases, prediction and parasitism as well as plants when it comes to seed dispersal and population. this paper seeks to look at the branches of ecology and bring out the key importance of ecology in the life of an ecosystem.

The study and research on insects have created a very critical development of the fundamental principles surrounding ecology: for example, population dynamics, predator- prey interactions, island biogeography, the evolution of social organizations, competition, as well as metapopulation ecology just to mention but a few. Significant economic and medical importance for a lot of insect species is one of the reasons for the coming up of entomological programs here in different institutions including government agencies and universities. There has been a stimulation and constant support for the study of the factors that influence insect abundance and its effects due to damages incurred by the agricultural crop industry as well as the transmission of human and livestock diseases (Schowalter, 2006).

8.2 ECOSYSTEM ECOLOGY

As earlier mentioned, an ecosystem is one of the most important units of an ecological organization and has very vague definitions of its boundaries. The same way species are interconnected within an ecosystem, ecosystems are interconnected also. The imports in one ecosystem are considered to be exports from another ecosystem and some of these features are; organic matter, nutrients, water, as well as energy, and some of these factors coming from terrestrial ecosystems are very important resources for countless aquatic ecosystems. The erosion of organic matter and nutrients by the wind from arid ecosystems is followed by a filtration process by the ecosystem downwind from the airstream. There is an overlap ingredient found in abiotic conditions that bring up the template which limits the number of options when it comes to community development, though already developed and established communities have the ability to modify and design abiotic conditions in an array of degrees (Prakash, 2008).

Ecosystems are considered to be complex systems that have a structure which is represented by abiotic resources in combination with different component species accompanied by their products and functions all existing among biotic and abiotic components of the same ecosystem. Bearing this in mind, this complexity goes all the way to the spatial depiction of various ecosystems. When it comes looking into to temporals and taxonomic, spatial complexity is one area that poses a serious challenge to ecologists that have to determine and evaluate the amount of complexity that can be safely ignored. Evolutionary and ecosystem ecologists have considered looking at different approaches towards dealing with special complexity and the evolutionary approach seems to emphasize more on adaptive aspects of the history of life, dynamics of population, as well as species interaction. This approach has hindered complexity studies to the interaction of very few insect species and their hosts, they are predators, as well as other environmental factors both biotic and abiotic which more often than not shrugs the complex feedback given by the ecosystem level. When it comes to the ecosystem approach, this system emphasizes more on the rates and directions of energy and matter and it restricts complexity study to fluxes found in functional groups and more often than not ignores the contribution given by individual species. Both of these approaches hinder the ability to completely fathom the feedbacks among population, individual, community, as well as ecosystem parameters which predicts a lot of effects that challenge the global environment surrounding these feedbacks (Schowalter, 2006).

8.3 ENVIRONMENTAL DISTURBANCE AND CHANGE

Environmental changes across spatial gradients and temporal gradients are considered very important components of an existing functional ecosystem approach when it comes to insect ecology. Insects are considered to be highly responsive to changes in the environment that at times have very fatal consequences. Some insects have an immense capability to travel long distance that allows them to locate and colonize resources that are isolated in those areas and other species of insects which are unable to fly, are considered vulnerable to the changes of the environment and many at times succumb to them. Due to their short life span and size, reproduction rates increase significantly, and several species change their order of magnitude based on seasons or annual time frame which brings down the time lag between changes in the environment on the adjustment of the population to the new conditions (Shields, 2017).

The main importance of environmental change and disturbance represented as a central theme in the ecology of insects comes out and husband recognized just recently. Environmental disturbance in particular you seem to bring forth a lot of understanding which helps predict individual adaptations of insects, the strategies of population, organization and the succession of various community types as well as the rates and regulations present in the processes of an ecosystem.

These changes in the environment or disturbances in the environment kill and affect individual species or insects individually which then affects their daily activities as well as their reproduction processes. There are very many species of insects that are reduced to only existing in various localities and some have been exposed to consequences such as local extinction. Interactions with other species as well as population strategies are greatly affected in an ecosystem in a variety of ways that contributed to the increase of more disturbance or in some cases mitigate changes in the environment and favor persistence in species that are less tolerant to change. This very aspect when it comes to insect ecology has a very important role and bears significant consequences for the ecosystem responses when it comes to global changes that result from coming through for anthropogenic activities. Some of these disturbances are not very permanent and considered to be temporary with only temporary consequences and at times no consequences at all bared by the ecosystem. There is however a possibility for major disturbances or environmental changes such as fires, flooding, insect outbreaks and many more which have detrimental consequences to the ecosystem, that these disturbances happened in as well as neighboring ecosystems (Schowalter, 2006).

8.4 ECOLOGY OF INDIVIDUAL INSECTS

In insect ecology, an individual organism is very important because it interacts with its environment and affects the ecosystem processes significantly via their joint physiological and behavioral feedbacks when it comes to environmental variations. The success of an individual insect is to find and make use of needed habitats and resources found in the environment to acquire reproductive advantages which is a great determiner of fitness and insects have various general characteristics which contribute to their ecological successes.

- The first part is the small size insects which are classified together with invertebrates and microorganisms that has allowed the consumption and use or food resources and habitats at a microscopic scale.
- The second part is the exoskeleton which protects against enemies of all kinds be it a predator or a property such as water or soil and this more often than not waterlogged which is very important for small organisms to be very flexible. The exercise has various advantages, but it also has a disadvantage of limiting the size

which is attainable by the insect and also matters of weight are limited well the same exoskeleton.

- The third part is the metamorphosis which is important for the groove of the insect and allows the division of habitat and resources that the insect is exposed to throughout its lifespan. Interspecies competition is greatly reduced by the fact that mature and adult insects have a difference in form and function causing them to live in different habitats and feed on completely different resources.
- The last part looks at the great advantage bestowed upon insects due to flight evolving first among them and it allows movement in short and long distances which opens up sources of new resources for the insects as well as an escape in the event of threats from predator or hush unfavorable conditions for the insect.

8.5 RESPONSES GIVEN IN THE FACE OF ABIOTIC CONDITIONS

8.5.1 Biomes

Biomes are considered to be global patterns of precipitation and temperature which reflect the interactions of latitudes, the global atmospheric and oceanic circulation patterns and the topography which establishes a regional template that shows physical results. All these characteristic support communities which are in short known as biomes. Tropical, subtropical, temperate, and the Arctic zones are all defined by the latitudinal gradient that is generated temperature-wise from the earth's equator to its poles. Humid and warm air rises in the tropics which draws air from higher latitudes to the equatorial merging point. The cool dried air flows downwards from the Equatorial zone and it gradually warms up as it flows downwards into the latitudes which are positioned at 30 degrees North and South. Latitudinal gradient stats are involved with this process, play an important role in the restriction of distribution organisms based on the tolerance ranges in terms of temperature and moisture. We have a lot of interaction with mountain ranges and Oceanic and atmospheric patterns which modify the latitudinal patterns present in temperature and precipitation. The mountains are responsible for forcing airflow upwards which causes cooling, the process of condensation and precipitation present in the windward side. This effect leads to the development of mesic environments data found on the windward side as well as arid environments such as the Leeward side of mountain ranges (Walter, 2005).

Very significant and relative the instinct hand in hand combination of temperature and precipitation is responsible for the very assembling on the dispersal of species of insects that are capable of surviving and also defining various characteristics when it comes to different community types of insects. Conditions of habitats in terrestrial biomes are greatly influenced by topographic relief, the substrate chemistry and structure as well as the wind exposure. Differences that are experienced in substrate chemistry and structure contribute to the limitation of survival for the existing plants and animals as well as characteristics of the surrounding biomes (Walter, 2005).

Sandy soils and loom soils are considered to be more fertile and are more prone to excavation more than others including ballast flows and serpentine soils that require the introduction of special adaptations for several plants and animals to survive. Insects that are found in windy habitats are considered to be more often than not flightless which is another petition that comes about and natural selection against being blown away during flight by strong winds. Aquatic biomes come about through topographic depressions which create zones for flowing and standing water. Existing insects in deeper zones in aquatic habitats exhibit very little temperature variations which allow the insects to continue their development and their normal activities all through the year including temperate regions. They are able to develop modifications and adaptations that allow them to settle in unless the chemistry and substrate structure of the habitat present in aquatic biomes is too unfavorable for them to exist.

Figure 8.1: Examples of various ecosystem structures in terrestrial biomes. Picture 1 represents tundra, picture 2 represents desert shrubland, Picture 3 represents the grassland, and picture 4 represents the tropical savanna.

[Source: Https://html1-f.scribdassets.com/7a58qtsocg41rqe7/images/34-2968056f18.jpg]

8.5.2 Environmental Variations

Various seasons in most biomes have a lot of different physical conditions. In temperate Ecosystems, they are characterized by obvious seasons of different temperatures some being cooler winters and an addition of warmer summers as well as a very significant difference in the precipitation patterns Present in different seasons all through the year. Physical conditions also differ via time due to the irregularity of events. Some changes in global circulation patterns also have effects biomes throughout the globe. one example is during the El Nino years which were 1982-1983 and 1997- 1998, there were extreme drought outbreaks experienced in the tropical regions as well as the strong storms that affected the functioning of various ecosystems found on higher latitudes. Many insects are very sensitive to the changes that take place in the environment including temperature and moisture which are a result of environmental variations and changes. There is also the inclusion of solar activity that is responsible for typical climatic conditions. when it comes to terrestrial habitats, they are very sensitive to the changes that occur in air temperature, the speeds of wind, the humidity present in these habitats, as well as other environmental conditions. in the case of Aquatic habitats, they are considered to be resistant to sudden changes in the environment as compared to other habitats, though they are very sensitive when it comes to change in rate flow, chemistry and depth, and pH fluctuations as well as gas concentration and pollution (Schowalter, 2006).

The physiological tolerance of insects is reflected by the physical conditions present in biomes that they call their habitat. Areas that have very high proportions of impervious surfaces such as roads parking lots and roofs slowly change the conditions of both aquatic and terrestrial ecosystems through an increase in albedo and precipitation runoff. The insects that are related to tundra biomes tend to have a lower tolerance for temperature ranges as compared to those insects that are related to tropical biomes. Higher thresholds of temperatures for the survival of the insects present in the tundra biomes could be the lower threshold for the survival of those insects found in tropical biomes.

Figure 8.2: An example of the aquatic biomes, picture 1 shows a stream, picture 2 shows a swamp and picture 3 shows a beaver pond.

[Source: https://html1-f.scribdassets.com/7a58qtsocg41rqe7/images/38-5677932ff2.jpg]

8.5.3 Disturbances

Disturbances are present in biomes and ecosystems in various ways and they take on different forms. They are caused by different reasons as well. Biotic and abiotic factors within biomes influence the patterns of disturbances and relatively discrete events that change the ecosystem conditions creating finer-scale landscape patches that have different disturbances and different histories in recovery. Some disturbances for example fires, storms, famines, droughts, flooding, among many others, change the vertical and the horizontal gradients present in temperature as well as moisture and the chemistry of water, which greatly affects the abiotic and biotic conditions that these insects are exposed to. Insects respond to these disturbances by developing adaptations that fit the disturbances that they are faced with (Prakash, 2008).

Figure 8.3: A representation of natural disturbances that take place in areas around the world in different environments. Picture 1 shows a fire that usually takes place in the savanna and grasslands. Picture 2 shows a storm that happened in a forested area. Picture 3 shows flooding.

[Source: Https://html2-f.scribdassets.com/7a58qtsocg41rqe7/images/42-3af-ba24f33.jpg]

There are however different anthropogenic Disturbances that are considered unique. Aquatic organisms have had a history of being exposed to minerals of different synthetic toxins that were introduced in the aquatic world. Natural disasters such as fires and many others do not generate a huge number of stamps inside that show exposed areas and underground root systems. The effects that are faced after search changes are more often than not very difficult to predict based on adaptations towards natural disasters and it can be experienced for long periods because of the local mechanisms that show a lock of reversal in the event extreme alterations in the substrate, the water conditions or vegetation happen. Disturbances can be considered to come in an array of ways, intensity and severity. Taking an example of low- intensity fires, they greatly affect the surface-dwelling organisms That could have developed adaptations against these kinds of disturbances, and an additional example is high- intensity fires pose a very serious threat to large proportions of the community. Winds and hurricanes pause very serious threats to forests which are considered to be serious dangers for many arthropods. All these different disturbances pause one thread or another when it comes to the survival, the reproduction processes, distribution on the general thriving of individual insects or insects that dwell and exist in communities (Shields, 2017).

8.5.4 Surviving variable Abiotic Conditions

Insects are considered to be very vulnerable particularly to changes in temperature, the availability of water and air, as well as air or water chemistry due to the fact that they have a large surface area ratio to volume. That being said, a lot of insects can coexist and thrive within suitable microsites that insulate the exposure to changes in the environment. Insects that live in an aquatic environment or in very deep soil or woody habitats are considered to be protected and shielded from the many changes of the environment especially in air temperature and relative humidity because of the high moisture content found in soil which mitigates heat penetration (Leather, 2008).

Due to the high number of insects that are considered to be vulnerable to environmental changes, the undergo a lot of stresses that include maintaining optimum body temperature, Water content and chemical processes which are a serious challenge in relation to survival when they are living in vulnerable habitats. Behavior shows a more malleable means of giving feedback when it comes to environmental variations which in comparison to physiology is a better way of responding to environmental variations because animals respond best to sensory signals when they feel threatened or are exposed to a dangerous environment. Mobile insects have an added advantage when it comes to mitigating exposure to environmental changes such as temperatures, availability of water or chemical conditions in comparison to insects that have limited mobility who more often than not suffer grief consequences when exposed to environmental changes. This applies also too many small insects that are flightless and are found on the soil surface is or just a few millimeters below the soil profile. Insects have developed adaptations that help them when it comes to surviving different environmental variations such as regulating their body temperatures, balancing their water content, changing their behavioral mechanisms and many more adaptations. The following are some of the adaptations adopted by insects to favor their survival when faced with environmental changes (Leather, 2008).

Thermoregulation: This is considered to be a process that insects maintain body temperature when they are in certain boundaries. Insects always could change or modify their body temperatures depending on ambient temperature and different situations which means that they are organisms whose body temperature depends on external influences and it also means that they do not have a stable internal body temperature. Insect display characteristic range in temperatures That are applicable in specific activities. We find that both terrestrial and aquatic insects are responsive. When it comes to accumulating thermal units. Significant alterations of thermal conditions are caused by anthropogenic conditions especially, in aquatic habitats. The dispensation of heated water, diversions, regulated water levels on the flows, an artificial combination of thermal strata as well as the canopy opening in riparian zones present in harvest and grazing, thoroughly changing the thermal environment it's of the aquatic species and fcvcr hcat- tolerant individuals on species over heat- intolerant individuals and species.

Some insects implement this adaptation by heat loss mechanisms such as harnessing convictions, Heat radiations, evaporative cooling, gaining or heat through warming up by shivering, warming up by basking in the sun, changing color, the ability to activate insulation just to mention but a few.

Water balance: Maintaining a homeostatic What are balance is considered to be a challenge for organisms that have high ratios in surface area to volume. The exoskeleton of an arthropod is very important in this mechanism for the control of water loss and water gain for the insect. Some arthropods have thicker cuticles and have the ability to secrete more waxes which prevents water loss which is an adaptation that insects in mesic environments do not have. Cuticula lipids that have higher melting points are considered to be less permeable to water loss when you compare them with lipids that have lower melting points.

Figure 8.4: This shows a tent caterpillar. This tent created by the caterpillar helps in reducing airflow as well as changes in temperature inside the tent itself regardless of the external temperatures.

[Source: http://www.wildflorida.com/wildlife/insects/Eastern_Tent_Caterpillar.php]

This is done by insects as a way of regulating the level of water inside of the insect as well as preventing water loss in the event unfair of environmental disasters or changes in temperatures especially. They adopt different textures, colors in their exoskeleton which helps in the conservation of water. As depicted in Figure 8.4 above, they can also construct structures or find habitats that take into consideration the conservation and balance of water. Having to deal with excess water is also an issue that insects have to deal with. This is mainly done through excretion through their skin among other ways and they make sure that they maximize the conservation of the nutrients which are consumed by the insect (Walter, 2005).

8.5.5 Air and Water Chemistry

This mainly affects insect physiology. The supply of oxygen is very important to survival, but it can also be very limited when it comes to different circumstances. Dissolved chemicals that are present in the air have very serious effects on the insect's respiration and development. The pH of soil or water has an effect on the exoskeletons functioning as well as other physiological processes. The concentration of different chemicals particularly those that are a result of industrial activity also affects insects (Walter, 2005).

8.6 FACTORS THAT AFFECT INSECT DISPERSAL BEHAVIOR

Insects have the ability to run away from unfavorable conditions of the environment and they go out looking for conducive conditions within Favorable temperatures, conducive moisture levels as well as chemical gradients all across watersheds and landscapes present in the environment. This is what is known as dispersal and it is the movement of insects from their original source or habitat in search of better living conditions due to various reasons, one of them being environmental variations. Some of the factors that affect this process are;

- Life history strategy.
- Crowding.
- Nutritional status.
- Habitat and resource conditions.
- Mechanism of dispersal.

8.6.1 Resource Acquisition

Resources in relation to insect ecology represent the presence of the factors that are critical to the survival and the growth of insect species mainly food and water as well as favorable environmental conditions. When you refer to the second law of Thermodynamics, it is shown that organisms gain energy for the collection of resources and they synthesize the molecules that are very important for life processes as well as growth and reproduction of the insect species. We have organic resources that are used by insects and they are different in many ways especially in quality, accessibility, as well as availability. This section of the chapter looks at the acquisition of these resources by the insect species as well as their effectiveness in different developmental and environmental conditions.

8.6.2 Resource Quality

When it comes to recourse quality, it is defined as the net energy and the nutritional value present in the food resources accounted for by the digestive ability of the individual insect. The nutritional value as well as energy present in organic molecules is as a result of the number, composition of elements, as well as the bonding energy of the consistent atoms. There is however evidence of lack of digestion of some organic resources and the bodily use of these organic resources under their components in the insects' body. That being said various plant tissues have a composition of lining and cellulose, which is only digestible by certain organisms, which makes it very hard for all species of insects to partake in various food resources. There is also a very serious challenge due to the toxins released by plants as a food resource because plants also have adaptations against animals and insects as well as man (Shields, 2017). These factors make it very hard for organisms to acquire resources especially in terms of food and it also compromises the quality of the food. Some of the plant chemical defenses are;

- Non- nitrogenous defenses.
- Nitrogenous defenses.
- Elemental defenses.

Even faced with these challenges, insects have the impeccable adaptation to protect themselves from internal parasites through ingesting antibiotics in plants that they eat, a gut modification that prevents penetration of the pathogens and cellular immunity against parasites.

8.6.3 Resource Acceptability

Here, in an ecosystem, we are informed that the array of resources as well as their biochemical and physical properties which include defensive mechanisms, stands too great and there is no way any species could exploit all the resources in that particular ecosystem. The feeding preferences of insects is determined by the particular physiological and behavioral adaptations that they have been able to obtain in the ecosystem which helps them get enough nutrients as well as being able to escape the toxins that are exposed to them by the feeding materials that they have found in the ecosystem as well as undigested materials. This all comes down in the decision making for what resources they can and will explore in their habitat. The resource acceptability shows the willingness for the insects to feed and also the given probability of being able to locate other suitable resources for the insects. Insects are provided with very little or limited time as well as energy to gather resources and they have also very limited energy spent in searching for their foods (Shields, 2017).

There is also a key factor that shows that insects feed on different resources and require different resources during different times and stages of their development. This all goes deeper to explain dispersal methods of insects as well as their selection when it comes to habitats. Lepidoptera members mostly feed on the roots, stems and foliage as larvae and they also feed on nectar when they develop into adults. This case also applies to cerambycid beetles as they are feeders of wood as larvae, but they feed on nectar and or pollen when they develop into adults. This whole concept of resource accessibility pushed the insect species to develop adaptations to survive and thrive in selective pressures offered by the habitats that they live in or the sources of the foods they require to develop and survive (Prakash, 2008).

8.6.4 Resource Availability

This means the distribution, the abundance and the apparency of the available accessible resources that the insects need for survival and development. The resources present in an ecosystem are considered to be available when they are distributed evenly throughout the ecosystem without limiting the densities or the concentration of the resources. The organisms that are present in the habitat often have to spread out to locate new resources that are considered to be sedentary. That being said, the available resources that are favorite and acceptable to the organisms of insect species are often finished or consumed in high amounts in the event the reproduction rate of the insects are above normal. This leads the insects to settle in places and ecosystems that have resources in very high concentration, or the balance of the available acceptable resources is at a really good balance. The insect species develop behaviors and physiological mechanisms to be able to locate, move towards food sources as well as orient themselves towards the concentrations of the food. This is done in the following ways:

- Foraging strategies.
- Orientation; whose cues involve the following steps,
- Information progressing.
- Responses to cues.
- Attraction of conspecific insects.
- Learningthat is a result of experiences of food gathering and locating which increases their efficiency.

8.7 POPULATION ECOLOGY

A population is defined as a group of species that are interbreeding with each other. The characteristics of a population are the attributes of structure for example density, age, sex the pattern of dispersion as well as the genetic composition which goes through several changes through the passing of space and time which is a response by the species to the changes in the environment. The structure of the insect's population and the dynamics of that has been as a result of research in ecology. This type and status of the organization of ecology have focused on ecological genetics, evolutionary ecology, biogeography, the development of the sampling method, management of pests, as well as the recovery of the insect species that are considered to be endangered. All these studies and branches of insect ecology have helped greatly in the continuous understanding of the phenomena at the population level.

The abundance of a lot of the insect species have effects on the order of magnitude based on their short time scales due to the fact that they have small sizes and reproductive rates that occur very rapidly. Changes that occur in that form of rapidness and dynamism affect the abundance in response the environmental changes that are more often than not stable and they facilitate the statistical evolution of population responses to the environmental factors hence making insects capable of colonizing new habitats as they exploit the conditions that favor them as well as taking advantage of the resources available for them quickly. That being said, the fact that they have short lifespans, small sizes as well as their dependence on chemical communications that they use in locating mates at low densities, and this limits their persistence of local and small populations in this period which more often than not leads to local extinction of insect species (Schowalter, 2006).

The dynamics of population often show the net effects of the diversity in the individual insects in their behavioral and physiological ways of interacting with the environment that they are exposed to. The improvement of individual insect success when it comes to the finding of mates, exploitation of recourses, reproduction, as well as the insect's ability to avoid mortality agents that determine the number of individual species, their spatial distribution, as well as the genetic composition of the insect species. When we talk of population structure, it simply means that it is a component present in the environment of the insects' population, and it provides important information that tells us about the individual insects' behavior and physiology leading to the fitness of the individual. The concept of population structure affects the structure of insect communities as well as the ecosystem processes. Every population is made up of the insect species as well as other members of the community, hence any changes that occur in the number of population of any species that is part of the community leads to detrimental effects on the population of which it feeds on, competes with or is the prey species (Schowalter, 2006). Aspects of population structure are as follows;

- Density of the insect species.
- Dispersion of the insect species.
- Metapopulation structure of the species.
- Age structure of the species.
- The sex ratio of the species.
- The genetic composition of the species as a result of the adaptations coming from the environmental changes and variations.
- Social insects.

In the steps involved for a population there are key aspects that determine the population as a result of variable reproduction of the species, the movement and the death of the individuals. The insect species play a big role in structuring and informing the three stages of population process which are;

- Natality which is the birth rate.
- Mortality is the death rate of the individual insects.
- Dispersal is defined as the movement of the individual insects in and out of the population.

The entire population system is mainly described in terms of the variables of the structure as well as the processes which give way to the change in the structure and these variables include;

- The Sex ratio.
- Density.
- The pattern of dispersion of the individual insect species and the demes.
- The genetic composition of the individual species of insects.
- The age structure of the insect species.

Some of the insect populations have been found to genetically modify themselves in very short spans of time all as a response to the short generation times that the species live as well as the very high reproductive rates that the species have. Over the years there have been huge technological advancements that allow researchers to study these patterns and in general, the population systems of the insects and the factors that surround the biogenetics that is related to their selection of habitats and this has a direct effect on the genetics of the species. This is in return an effect faced by the ecosystem as the communities that are in coexistence with the insect species have to adjust to the environment or they at times disperse.

There is the concept of population fluctuation. This comes about due to several factors that are present in the ecosystem as well as the environment. Environmental variations, seasons of the year, changes in population abundance of other species in the ecosystem, changes in the resource abundance, acceptability or accessibility, just to mention a few, are all key factors that could lead to the insect species population fluctuation in a given area of an ecosystem. Factors that affect the population size of insects are;

- Density independent factors that modify the population size of the species.
- Density- dependent factor which regulates the population of the species.
- Regulatory mechanisms.

The fluctuation of insects is attributed also to the ecosystem patterns and processes and also the level of crowding in a given area of the members of any one population. The increase in population can have effects on the resources available which leads to higher consumption of food, reduction in natality, an increase in mortality and an increase in dispersal which reduces density all at ones. The population could be regulated on a top-bottom basis which is the resource limitation factor approach or a bottom-top basis which is a predation factor approach. This study has given way to the development of population dynamic models which helps in interpreting the changes in abundance of insects as well as the effects that are faced by crops, the ranges they occupy and the resources of the forest (Prakash, 2008). Some of the models are;

- General models that comprise of the Verhulst pearl equation which is a combination of the population size; per capita natality, mortality, and dispersal; and the environment carrying capacity.

- The logistic model whose equation is aimed at explaining the sigmoid curve reaching an asymptote at carrying capacity.
- The complex model.
- The model evaluation.
- The computerized models.

There are very few models developed in directed to be able to predict the effects felt by the population dynamics for the ecosystem processes with an exception of commodity production.

8.8 BIOGEOGRAPHY

This is defined as the study of the distribution of insect species in and ecosystem in any geographic area in geological time. The geographic ranges of the species usually happen and show the ability of the species to tolerate the geographic gradients presented in the physical conditions of the environment. That being said, most insect species do not entirely occupy the habitats they are in and find to have suitable living conditions for them due to the discontinuity presented by barriers and certain disturbances.

Geographic distribution of the species population can be explained on several scales, and the largest scale states that some species have population distributions that spread out in large areas of the world and the smaller scale states that insect species individually are likely to exist in suitable areas of settlement across only a portion of a biome or a landscape at largest. The distribution of individual species geographically has a chance of changing due to changes that are present in the conditions of the habitat or the dispersal. Factors that affect population distribution of insects geographically has been a serious attraction for ecologists over the last 200 years. By distribution, this means the scale difference geographically. There are six different floral and faunal biogeographical realms that can be identified, and this confirms the history and the movement in the plates of the earth's tectonics (Prakash, 2008).

When it comes to rapid population growth, this always leads to larger numbers of individual insects of a population dispersing and this increases and maximizes the colonization of new grounds for the species to develop on. The decline in the population of the insects creates an isolation refuge which maintains the demes of the metapopulation and this level of movement of the individual species of insects is the key determining factor of genetic heterogeneity as well as the insect's ability to recolonize the patches again as a result of local extinction (Shields, 2017).

8.8.1 Community Ecology

The co-occurrence of species in a given area has an interaction to a certain degree on a direct and an indirect level. These interactions can be explained as mechanisms that are aimed to control the population dynamics of the area; they co-exist in. this gives rise to community structure, the control of energy rates as well as matter fluxes, in the long run developing a functioning ecosystem. A community comprises animals, plants and microbial species in a given area and these communities have changed over time as a result of responding to the environmental changes that are faced by the area. The interactions of species are different in an array of ways though they produce a combination of positive, neutral and negative effects for all species present in the community. The classes of interactions of the species present include;

- Competition through exploitation which comes as a result of resource acquisition and recourse protection by species.
- Predation which comes about by the species in the area being prey and others being the predator.
- Symbiosis is the association of the host species and the symbiotic species which includes parasitism, commensalism and mutualism.

The interactions are affected in various ways which influence their nature and their intensity. Some of the factors that affect the interactions are:

- Abiotic factors affect the competitively if the species, the predatory or the ability to defend themselves.
- Resource availability.
- Indirect effects of other species.

The interactions between the species affecting a given species are likely to preserve the population size in a narrow range. There are measures in place to stabilize the interactions of species in order to avoid local extinction happening in multiple areas as a way pf preserving species of all kinds and encouraging interactions between species. This helps maintain a good balance in the population numbers as well as helps the species adapt to one another's presence. This is called the species diversity that is described in 2 ways;

- Richness is the number of species in any given community.
- Evenness which is the measurement of relative abundance.

Habitat stability, the availability of resources as well as species interacting with one another is a key factor in developing a stable community structure.

8.8.2 Ecosystem Level

This is the integration of species interactions as well as the community structure and the species response to and effects on an abiotic environment.

Figure 8.5: An image of a community of aquatic organisms which is inclusive of microflora as well as invertebrates. The plant in the image develops structures to hold water. This is a representation of the small-scale ecosystem that has measurable energy inputs as well as matter.

[Source: https://html2-f.scribdassets.com/7a58qtsocg41rqe7/images/329-fba-ce6a4e6.jpg]

Ecosystems are all characterized by their functionality and their structure which is a measure of the organization of the ecosystem and the biological modifications and adaptations of the abiotic conditions. An ecosystem structure refers to the sources and sinks of energy and matter as well as their relationships towards one another which determines the capacity of the ecosystem to buffer. Some of the structures of an ecosystem are; a Trophic structure and a Spatial variability. The abiotic pools are oceans, the atmosphere and the sediments which reflect the energy and matter for use. The biotic pools are comprised of different organisms in the species community such as autotrophs and heterotrophs. The available energy in an ecosystem is derived mainly from solar radiations, the captured and stored carbohydrates by autotrophs via photosynthesis (Walter, 2005).

The rate of energy in total is determined by the level of exposure to the sun, water availability and biomass. The currency that the organisms gain and concentrate material resources needed for their development, as well as reproduction, is mainly always energy. Ecosystems also have the ability to adjust local and regional conditions of climate which is seen in the control of soil warming, evaporation, soil erosion and the speeds of the wind which is all from the density as well as vertical architecture. The organisms will affect the structure of the forest canopy as well as the atmospheric interactions. The next step is to look at herbivory. This is the rate of consumption of the foliage, roots, stems, flowers, fruits and seeds of plants by any animals or organisms. This is seen to be a key element of ecosystem processes to reduce the density of the plants available in the ecosystem, the transfer of nutrients and mass into the soil or the water which affects the habitat as well as the condition of the resources for the organisms. The primary herbivores in any given habitat covered with plants or vegetation of any kind are insects (Walter, 2005).

Figure 8.6: An image reflecting the consumption of plants by insects that increases the canopy porosity.

[Source: https://www.sciencedirect.com/topics/agricultural-and-biological-sciences/folivore]

The effects of herbivory are all dependent on the feeding density and intensity of the organisms. The different types of herbivory have different effects on tissues, the production, the translocation as well as the accumulation of different degrees of photosynthesis. There are a number of methods that have been developed in order to measure the effects and the different intensities of herbivory. One of the most commonly used methods is the estimation of the consumption rates done by the individual species and the effects translated to their population number. Another method is measuring through finding out the level of missing plant biomass. The entire process of evaluating the effects of herbivory is inclusive of looking into different arrays of plants and different responses by the ecosystem. There are very few studies that have ventured into the different effects faced by the severity of herbivory and neither has there been enough of them to answer questions on matters of the biochemical cycle of insect herbivores or abiotic conditions. there are however proven effects of herbivory which are;

- Reduction in vegetation.
- Light penetration in vegetative areas.
- Precipitation.
- Wind speeds.
- Soil warming and erosion.
- Fluctuations in relative humidity.
- Transpiration.

All these effects have a ripple effect on the climate and there is an increased probability of an increase in disturbances in the future.

8.8.3 Pollination, Seed Dispersal and Predation

These are all very important processes that are carried out by insects as well as animals and there is a direct effect on reproduction rates and distribution of plants. Insects are the main agents of pollination and seed dispersal as well as seed predation in any given ecosystem. There are different types and patterns of pollination carried out by insects. Some examples are; Pollinator functional groups that are restricted to groups of insects which have a group of functional insects that include thrips, flies, beetles ,etc. these are known as the generalist pollinators. The specialist pollinators include birds mostly and they exploit particular flora. Pollen feeders are insects and animals that feed mainly on pollen produced by flowering plants (Leather, 2008).

Figure 8.7: This is a depiction representing the process of pollination. Picture A shows a honeybee and picture B shows the scarab beetle.

[Source: https://html2-f.scribdassets.com/7a58qtsocg41rqe7/images/398-602f752e91.jpg]

Seed predators are those that are known to consume the entire reproductive organ of flowering plants which is the whole flower. In pollination, the insects and birds move from one plant to another which assists in cross-pollination as well as self-pollination. Seed dispersal is very important for the species of all plants where these agents normally insects, birds and animals help in relocation and spreading of the plant species which assists in the development and shows successful results when it comes to reproduction evaluation. Seed predation and dispersal are also carried out in functional groups just like pollination.

In this light, we should acknowledge the fact that insects are agents of regulation of the processes that take place in an ecosystem. They affect environments that they are around through temporal and spatial patterns in the form of the acquisition of resources as well as redistribution of these very resources. The aspect of insects regulating the ecosystem processes is undoubtedly one of the most significant matters that stand out and reflect this regulation as a method of conservation of the ecosystem by the insects. There has been debate on whether or not group selection has any effect on the direct selection of the individual species attributes. Individual attributes have an effect on organisms in the same environment as the individual as well as the environmental conditions and these effects bring forth feedback which tells us the general fitness of an individual (Leather, 2008).

8.8.4 Application

When we talk about insect ecology, this is seen to be the direct connecting point in between the understanding of the interaction of insects and their environments and the needed pest maintenance application, restoration of the ecosystem and development of many different key aspects of management of different ecosystems all over the world. When we understand the ever-growing and developing relationship between the environment and the insects, this gives light to the deeper synthesis of insects within the context of an ecosystem as well as global processes. The only reason insects are viewed as pests is when their population in any given ecosystem or area has a serious effect on the crop present and the insects' capability in spreading of diseases in humans and animals. The outbreaks usually occur when there are favorable environmental conditions for the insects to exist and thrive which is usually attributed to natural disasters.

In the management of the crop, forest and urban pests, there have been major advancements where there has been better crop produced and there has been the reduction of animal and human diseases by the pests. This boosts the health of the crop produce, farm animal produce and the general health of human beings. Ecology has allowed the development of many different and effective tactics such as elimination of conducive conditions, enhanced defenses of plants, insect growth regulation, predators and parasites as well as pheromones which all contribute in an approach called the integrated pest maintenance. There has been the genetic engineering of plants and animals in response to the pests which is attributed to the defenses that these animals and plants get to be able to resist the pests. This has also helped in the devising of new agricultural methods of planting crops as a way of pest control as well as the additives and chemicals that help modify the soil not to favor pests (Schowalter, 2006).

When it comes to conservation and restoration of the ecosystem, ecology has shed light on insect species that are considered endangered species. Insects such as the American burying beetle and the Fender's blue butterfly are some of the insect species that are targeted by the agenda of conservation. This is covered under the Endangered Species Act of 1973 which was enacted in the United States of America with the goal of recovery and restoration of the species considered to be threatened or endangered and are facing near extinction. These species are created for a conservation area that is dedicated to restoring their declined population numbers and the study of ecology helps a great deal in being able to create a suitable environment that is needed by these species undergoing this process.

In the indicators of the environmental conditions, it has become a useful factor in the increase of the understanding of insect responses to the environment and its factors. This is due to the sensitivity towards climate and biochemical changes that happen in the resources as well as the high rates of reproduction that makes insects likely to warnings prior to the changes that are yet to be spotted in the plant abundance. Insects are known to indicate sighs that help in predicting the quality of water. There is a huge capability that insects have to change the conditions of the environment in very dramatic ways. They can alter the vegetation structure, the direction and rate of material flow and energy, and all this is through the ecosystem. This is called ecosystem engineering, and this takes place in order to balance the elements present in the ecosystem and stabilize life in the site (Schowalter, 2006).

There needs to be a deeper look and investigation in the ecological field in relation to insects which will bring in a lot of benefits but mainly, unlock and answer questions that are yet to be looked into, for example, being able to use the evidence present in the ecosystem to predict future effects of the current state of the ecosystem which will help in the development of strategies that are geared to uphold the environment for the greater good of all species be they animals, plants or humans or organisms. The testing of ecosystem hypotheses would entail an array of approaches that would look into the population dynamics and community dynamics as well. There are a few considerations that are viewed to be important;

- The experimental design demands the attention of the statistical independence of samples. This means that individuals in a particular population could be used as replacements in the place of population and community properties, and data must be pooled and evaluated at the ecosystem level of the ecosystem in question in order to examine and bounce the variables around to give the relevant information from the findings.

- The research has to look into the responses of insects to, effects on, conditions of the ecosystem also have to be looked at a bigger range and scale of ecosystem variables.
- Spatial and temporal scales of research and perspective have to be broadened and expounded adequately.
- Many ecosystem components have to be provided with special attention such as subterranean areas.
- The principles that are present in insect ecology have to be applied in the improvement of the management of the population of insects as well as the resources present in the ecosystems.

In conclusion, this section has explained the various attributes that are as a result of insect interaction with the environment. There has been a feature on the adaptations that have been taken up by plants and by the insects' species themselves in order to survive. There has also been and explanation as to why insects migrate from one ecosystem to another and that has also shed a light on the effects that are borne by the ecosystem that the insects have migrated from. The determining factors that dictate the settling in of the insect species in any given habitat and the ability to thrive is from mainly, the resource availability, the resource acceptability, the resource accessibility, and the ability of the insect species to colonize the new site that they have inhibited in the aim to develop, grow and increase their reproduction rates which in turn has an effect on the community present in the ecosystem. In the event there is a deficiency or a disturbance that could be a natural disaster or a manmade disaster such as fires, floods, earthquakes, landslides, nuclear activity, heavy rains and many more, the insects are forced to relocate and search for better places where the resources, as well as the conditions of the environment, favor them. There is also the element of colonization of new sites by the insects which creates and interacts with other species present in the site.

We have also looked at the population dynamics that are determined by the interactions of the insect species with the environment as well as their interactions with other species in any given ecosystem. There has been a key factor that brings out the insect species as restoration and conservation agents of the environment as well as their important role in the dispersal of seeds, the seed predation and the pollination processes of the flowering plants and also those plants that cannot flower but require pollination and seed dispersal. Above all, there is a deeper understanding that is brought out by ecological studies of insects that interpret the manner in which an ecosystem will shape itself to function and be structured. The functionality of the ecosystem boils down to the structure and the harmony of the processes that take place in the ecosystem in question.

Ecology of insects has shed light on many levels when it comes to the insect world in relation to the ecosystem and the environmental variations at large. There have been technological advancements that are very important in the lives of man especially pest control which was made simpler by the exploration and diving into insect ecology, though there is room for improvement and more information to be discovered. With the current times where we experience a lot of environmental variations and climatic changes that have detrimental effects on the crop produce that we have as humans and it reflects on the health and stability of the ecosystem, giving feedback on the individual species success. having a lot of funding and force-directed into insect ecology, there will be better ways to deal with climate change for the sake of preserving the environment because the insects will be a key element in the prediction of future changes or conditions in the environment.

This only stamps in the importance of insect ecology in the direction of understanding the ever -growing and developing relationship between insects and their environments. The application and incorporation of insect ecology in modern countries, as well as developing countries, will be an answer to many existing problems such as crop sustainability, water quality, pollution, erosion, desertification, diseases in plants, humans, and animals, as well as insect outbreaks such as locust plagues and many more. With modern science available, there would be more good done by simply understanding the interaction of insects with the environment as well as the interactions within themselves.

9

Insect Relationship to Humans

CONTENTS

9.1 INTRODUCTION

Insects are a dominant existing life form in nature and they have the largest and evidently the most diverse species in the class arthropod. There are about a million insect species that have been described by scientists and others are yet to be described and interestingly have an invasive nature and millions of insects can inhabit just an acre of land. Arthropods are characterized as having a hard exoskeleton, segmented and jointed legs. According to the scientific classification of living things they fall within the kingdom Animalia, phylum arthropod, subphylum hexapod and order insect. Arthropods are further divided into four classes this is inclusive of insects, crustaceans, myriapods and arachnids. Several insect species exist in our ecosystem and they have been proven to be the most productive and beneficial arthropod species which is a result of their ability to move and conquer new environments. They have the largest and the most diverse species in the class arthropod and have managed to conquer different environments but this is in exception of the polar region and oceanic which is a result of their ability to fly and the nature of their exoskeleton. Entomologists have done a lot of research to try and understand these different species in their own habitat including their life ecology, evolution, social behavior and insect's relationship with other animals, humans and their environment. The most recent classification of insects divides them into 29 broad orders this includes Dipthera, Mantodea, Blattodea, Coleoptera, Odanata, neuropteran, Embioptera, Zoraptera, Mecoptera, Megaloptera just to mention a few. From studying insects' entomologists have discovered the usefulness and importance of insects to the ecosystem and humans. They play various roles.

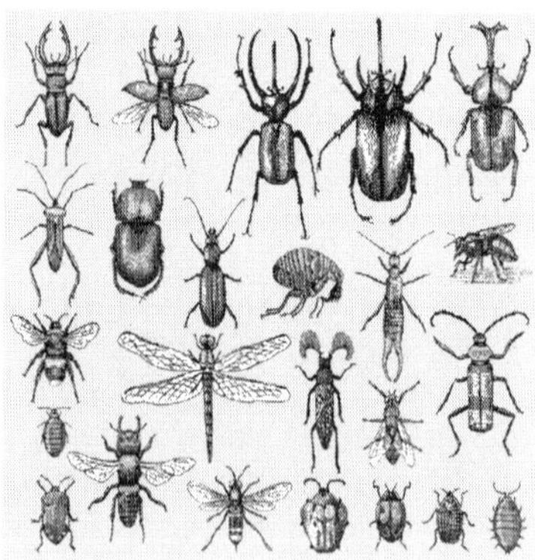

Figure 9.1: Different types of insects.

[Source: https://doi.org/10.1016/j.biocon.2019.01.020]

Insects have existed since time immemorial and can be referred to as ancient animal species despite this their evolutionary relationship is yet to be clearly understood. Their grouping with centipedes and millipedes despite their utter lack of resemblance can be attributed to their convergent adaptation although evidence has emerged linking them to crustaceans. There is a probability they existed before the Devonian and by the Carboniferous period. Their adaptation to fly was a good strategy within the Permian period where they achieved their significant diversity as no other arthropod group as attained the ability to fly.by this period a lot of insects had evolved in terms of their morphology. The most recently evolved order in the insect species is Hymenoptera which is inclusive of ants, bees and wasps and also Lepidoptera which includes moths and butterflies which appear in the Jurassic period as fossils. Eocene period was marked by the appearance of the Mantodea. During the Cenozoic period, the modern insect genus developed. Insects discovered during the said period were found preserved in amber and were in good condition. The morphology of such species was easy to make a comparison in relation to modern species and this type of study referred to as paleontology (Heming, 2018).

Insects can be studied for various reasons. Their ecologies vary and they can dominate food chains in large numbers. They also have different feeding techniques including the consumption of dead, living fungal mater, filter feeding, hcrbivorous nature, parasitism, being predators and even sap feeding. Insects can exist in a variety of settings including on land, water, soil. their interactions and lifestyle maybe be solitary, sub- social, gregarious or highly social. They may be active during the day or night or in some cases both. Their survival in a variety of environments and climatic settings can be attributed to their life cycles as insects. The relationship between insects and humans can only be described as complex. This relationship can categorically fit into three classes of a biotic relationship and these categories can be a biotic relationship. The association can be of parasitism, commensalism or be mutual. Depending on the nature of their relationship during their interactions at the time people decide their course of action on how to deal with the insect. This includes either exterminating them or encouraging them to flourish in the world for instance if an insect becomes a pest people will choose to get rid of them from their homes or their environment and if their presence controls the population of other pests then people will allow them to exist. Since antiquity insects have pestered our plants and animals, infected us with diseases, attacked our food crops and infested our food stores but they also have a considerable amount of benefits such as pollination, honey and wax production, beauty, biological control of pests and weeds activities which have been considered as just a fraction of how insects shape humankind perception of insects. We can make a conclusive inference that insects have well -defined advantages and disadvantages to humans and the environment. Still throughout history insects however have managed tap permeate the cultural, spiritual and scientific engagements (Heming, 2018).

Figure 9.2: Insects through the ages: (A) The largest insect ever, *Meganeuropsis permiana* (Protodonata), and Early Permian griffenfly. (B) A giant, stem-group flea, *Pseudopulex wangi* (Protosiphonaptera). (C) The earliest microphysid bug, *Popovophysa entzmingeri* (Hemiptera. (D) A solitary bee, *Oligochlora eickworti* (Hymenoptera), in Early Miocene amber from the Dominican Republic. (E) The nymphaline *Prodryas persephone* fossil (Lepidoptera), from the Eocene-Oligocene boundary of Florissant, Colorado, USA.

[Source: https://doi.org/10.1016/j.cub.2015.07.059]

Figure 9.3: Nature's inordinate fondness for six legs: Modern hexapod diversity representing the summation of over 400 million years of high speciation rates and low levels of natural extinction.

[Source: https://doi.org/10.1016/j.cub.2015.07.059]

9.2 INSECTS AS PESTS

Pests are animals or plants with a detrimental effect to humans or plants alike, forests and livestock these organisms can also be a nuisance. A huge percentage of insects are considered as pests by humans (an approximately 1%). These pests can further be classified as parasitic (for example lice and bedbugs), disease vectors or pathogenic insects (flies and mosquitoes) which result to diseases and human death, those that cause damage which includes termites and there is a group which is responsible for destroying agricultural goods such as weevils and locusts which are responsible for the loss due to damage. Many scientists are in the frontline in identifying different means of pest control and are therefore actively involved in research that sees the creation or the generation of various insecticides but has managed to rely on biological means that cause less destruction and disturbance to the environment. This includes the use of one population in an ecosystem to reduce another population. The use of insecticides over time has been associated with affecting other populations by killing animals that have been considered as beneficial to man and the environment (Morris, 2006).

9.2.1 Household Pests

Long before civilization insects fed on food for human consumption such as seeds, vegetables and fruits but with human advancement they started planting, storing and processing food this relationship changed and they were considered enemies by humans. Some of these insects also adapted and started inhabiting households as time went by. Their detrimental effects will be explained using examples that are relevant and relatable for instance the larvae of moths and beetles can be destructive to rugs, woolen clothing, and hides. Furniture, cabinets, floors and finished wooden products are destroyed by powder post beetles. Cereal products, grains, legumes, spices can be infested by the cigarette beetle. Candy, crackers, herbs, and legumes can be attacked by the drug store beetle. Cockroaches such as the American, oriental, German, and brown- banded species are known to invade food stores and destroy paper products and clothing. Ants have the frequency to invade homes in search of water and foods but especially sugary foods. Dried fruits, sugar, cured meats, nuts, pasta, bread, cereals can be infested by the omnivorous saw-toothed grain beetle. The most important pest that is invasive to food stored in supermarkets and homes is the confused flour beetle. The rice weevil invades rice while the green weevil prefers barley and wheat products. Coarsely ground grains are fed on by Indian meal moth larvae (Morris, 2006).

9.2.2 Pests that are Invasive to Domestic Animals

Domestic animals commonly kept by human beings include poultry, livestock and pets. Insects such as lice, flies and flies infect, weakens and kill these domestic animals. The insects damage hides and their skins. They also consequently lead to a reduction of animal production of meat, milk, egg and wool leading to huge economic losses. Species differ in invasion some attack a variety of species while others are specific to an individual species. Others even attack both wild and domestic species. Blood feeding pests include mosquitoes, horn flies, black flies, and screwworms. Stampedes among herds of livestock can be caused by female heel flies that have irritating bites. Blood loss leading to anemia may be caused by flies and horn flies. Injured animals with open wounds can be infested by screwworm maggots that live in these wounds, they can also feed on live tissues ultimately may cause death to the host animal. Anthrax and other pathogens that lead to mastitis and keratitis can be transmitted by blowflies, horseflies and stable flies. Irritation and scratching against posts and fences can be caused by cattle lice and as a result, there can be exposure of skin, hair and blood loss these losses can lead to failure of weight gain. Infectious diseases such as swine pox can be carried by hog lice. Other species such as beetles, fleas, roaches and beetles are intermediary hosts of disease- causing organisms such as tapeworm, poultry and heartworm and may infect animals if ingested (Morris, 2006).

Figure 9.4: Death of livestock as a result of pest.

[Source: https://www.agriforhealth.com/2018/12/introduction.html]

9.2.3 Agricultural Pests

Human beings cultivate plants to be used for human consumption and for economic gain. Plant cultivation has affected how humans perceive insects. Overtime with the practice of large scale agriculture farming has resulted in invasion by a vast number of insect species. Once these organisms invade and establish their presence they are able to multiply and spread and thus become an economic threat ruining crops for an entire region. Insects can invade. Agricultural pests such as thrips, butterflies, beetles, flies, moths, ants and homopterans can cause serious damage throughout the world. Other than causing huge economic losses to humans they can indirectly cause malnutrition to humans, starvation and death. Insects can reduce crop yields by the removal of fluids that are nutritious from crops while feeding on their vegetative structures. Other insects transfer plant pathogens such as bacteria and fungi hence reducing plant vigor. Beetles are known to spread viral diseases (Klowden, 2013).

Figure 9.5: Coffee green scale.

[Source: https://mrgoutham.blogspot.com/2011/06/coffee-insect-pests-pictures.html]

9.2.4 Forest Pests

Trees in the temperate forests, both dead and living trees are attacked by larval moths and beetles and to a lesser extent few flies, wasps and sawfly. Some insects bore into trunks while others attack seeds, buds, foliage and roots. Live trees suffer in the process as their vitality, growth is affected. The western pine beetle is found in the west of the United States of America and is of native origin and dark in color, it is known for its attack on coulter and ponderosa pine trees with a length diameter of greater than 15cm. this beetles attack trees without considering the age, vitality or health of these trees. Attacks and outbreaks also occur within species of mixed conifers yet the effect of the attack by these beetles on old trees remains elusive. During the yearly outbreaks which are often large over one million trees can be killed leading to very huge losses as the trees that would have been used for construction as timber are destroyed. The outbreaks can occur concurrently with wildfires or afterward (Klowden, 2013).

In unfortunate cases, these insects attack trees that are on the verge of stress and they fail victim to these pests. Other than economic damages insects also cause the loss of habitats for other animals and cause soil erosion. Other disasters can also cause the outbreak of pests.

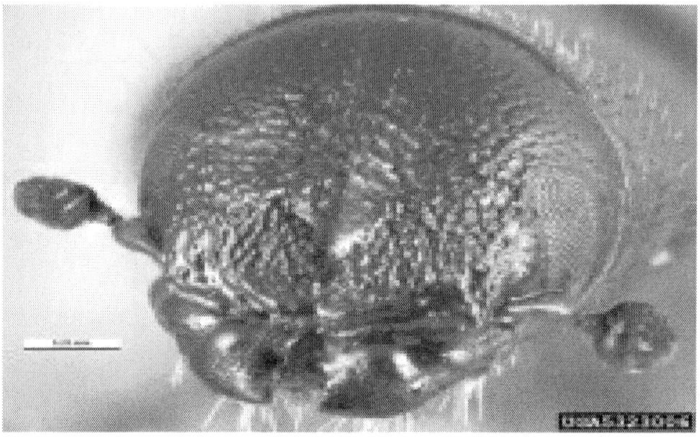

Figure 9.6: Western pine beetle.

[Source: https://www.forestpests.org/vd/23.html]

9.2.5 Insects as Vectors of Human Pathogens and Diseases

Important insect vectors of human disease include lice, mosquitoes, flies, fleas and assassin bugs and as a result, many people have died from infectious diseases as the cures to these diseases had not been found. Assassin bugs are carriers of protozoa that cause Chaga's disease while houseflies are involved in the spread of myiasis. Yellow fever is caused by mosquitoes and it is a viral infection other infections include dengue fever, eastern and western encephalitis. The female anopheles causes malaria. The tsetse flies cause nagana in animals and sleeping sickness in humans. Fleas are known to consume plague-causing bacteria from infected rodents and this is passed on to humans causing diseases such as bubonic, septicemic and pneumonic plagues. Lice are also other insects that are pathogenic but unlike mosquitoes, they do not transfer the pathogen instead humans will infect themselves with the pathogen by crushing them (Shields, 2017).

Figure 9.7: A mosquito on human skin.

[Source: https://lakenormanpest.com/insect-diseases-humans/]

9.2.6 Structural Pests

Building materials such as timber infest termites, wood-boring beetles and carpenter ants. Their damages are imposed on wood until it is completely hollowed out. Termites have been termed as the most destructive insects in relation to timber.

Figure 9.8: Termites as structural pests.

[Source:https://pestsd.com]

Pests can be controlled through various means including chemical, physical, cultural and biological means. How effective a pest control method is fully dependent on understanding the life cycle of the insects and correctly identifying the insect and the methods can be therefore being directed at the pests during the most vulnerable stages of their development. Chemical controls include sprays, dusts, aerosols, and fumigants they include oils, inorganics, botanicals and synthetic organophosphates. These insects function as poisons killing them through asphyxia and abrasion resulting in the loss of fluids within the body of the insects. Mechanical and physical controls include intensive methods such as trapping, handpicking, and the use of barriers that are really not practical when employed in a large piece of land. Cultural controls include the application of traditional farming methods such as crop rotation which is solely based on the knowledge of a pest's life cycle. Another control of insects is biological control which can include the use of one animal species to destroy another species (Berenbaum, 2010).

9.3 BENEFICIAL ROLES OF INSECTS

Economic entomology involves the study of the benefit that insects have to humankind that is economically driven. As seen previously insects have a collective number of destructive roles and contrary to popular opinion they have been proven to also have benefits to man and the entire ecosystem. The insects that are most beneficial are those that are reared for food or provide products that are used as food by people. Insects are also useful in plant reproduction as they assist in the pollination process.

Honey is one of the most significant products with economic significance to humans produced by bees hence beekeeping is a common commercial practice in some regions especially among tribes and communities that reside in the forest. It has been relied on as a major source of food and nutrition by these communities. Bees that provide honey can have significance in the pollination of crop species. Beeswax is also a product from bees that is used in the manufacture of candles, ointments, polishes and it is also used in the cosmetics industry to produce lipsticks, creams and lotions. Britain is known for importing huge amounts of beeswax (Morris, 2006).

Figure 9.9: Bees on a beehive producing honey.

[Source: https://www.teambonding.com/programs/beehive-building/]

Using wasps, bees, ants, and butterflies; insects are actively involved in the pollination of flowering plants and this has led to the development of a mutual relationship between the multiple species. Insects will get nectar from plants and in the process, the insects assist with the transfer of pollen grains from one plant to another. This relationship can be termed as involuntarily symbiotic. Through this process, they improve the probability of cross-pollination increased,therefore, improving improve the genetic and morphologically qualities of plants which is crucial to humans as this affects agriculture which is important in the food and commercial industry. Seed distribution by plants can be aided by ants as well as the pollination process. This role actively taken by ants becomes important in plant spread and eventually the achievement of diversity in plants thus making improvements to the plant ecosystem. With the problem of a decrease in pollination species, the pollinators can be bred or cultivated to positively impact on the pollination process by having an adequate and constant supply of pollinators and this practice is commonly employed in greenhouses and orchards (Marshall Cavendish, 2000).

Figure 9.10: A bee sucking nectar from a flower with visible pollen grains sticking on its appendages.

[Source: https://www.studyread.com/importance-of-insects/]

Silk is a product that is extracted from wild and bred caterpillars that later bloom into butterflies. Sericulture is an art that employs techniques that are useful to silkworm rearing and its production. Currently, other fabrics have replaced silk but it is still the material chosen for surgical procedures. Trade relationships between china and other countries were driven by silk as a product which was commonly produced in China.

Synthetic substitutes have replaced lac which was previously extracted from lac insects and is commonly cultivated in kerria lacca. The cultivation of lac is initiated when farmers acquire a stick carrying eggs that are ready to hatch and he or she continues and wraps it around a tree to be infested. The lac insects will then invade the branches of this host trees which produce a resinous pigment. These branches are cut and harvested and are referred to as sticklac which is crushed and filtered to remove unwanted residue. The resulting product seedlac is purified using solvent extraction to produce lac. This product is useful as a wood finish, a cosmetic, a dye, and traditionally as anti-obesity medicine and finally as violin varnish. Cochineal insects have a dye extracted from them which has now been substituted with time to due to technological advancement. Tannin is another dye that acquired from insect galls and it's known for the function of tanning hides in the ink production (Marshall Cavendish, 2000).

A balance in nature relies on predator and parasite relationships, a large number of which are from the insect species. Within the insect family, some are insectivorous which means they feed off of other insects and their importance is such that they can beneficial to humans in the biological control of insects that are considered as pests especially those that cause harm and losses in the agricultural sector and also to structures such houses and other equipment examples include aphids can feed on agricultural crops and are a problem to farmers but we have ladybugs which will feed on them and can be used by farmers to eliminate and reduce the aphid population. As much as other predators have taken the responsibility of reducing the insect population a significant percentage can be accounted for by insects feeding on other insects or feeding on other animals including vertebrates using ants as an example. Without other animals keeping insects in check, they can gain an explosive increase in population and continue causing harm [especially pests] (Shields, 2017).

Figure 9.11: Ants feeding on aphids from a plant.

[Source: https://www.gardeningknowhow.com]

Insects are also relevant in the field of medicine. Europeans in the early ages used ladybirds in powdered form to offer relief for people with toothaches and also as a cure for measles and colic. The Japanese also traditionally used different species of beetles to cure illnesses such as convulsions and hemorrhoids. Concoctions were also derived from insects in medieval Europe. The practice of using insects for medicinal purposes was adopted in different regions in the world and this has changed as a result of technological advancement and industrial growth (Khaliq et al, 2014).

In the modern age, some insects are being used for medicinal purposes for example the defense system of insects such as bees is exploited for pharmaceutical use. Bee stings have been used to treat patients with rheumatism to help in pain management. Cantharidin which is produced by blister beetles and Spanish flies is used in the treatment of epilepsy and asthma but can cause lethal side effects when taken in large amounts such as kidney failure.

Insects in their larvae stage were previously used by doctors in wound treatment to prevent gangrene which is a result of lack of blood supply which causes tissue and cell death within the human body. These insect life forms were used to eat up the dead flesh and it is still practiced in the modern age- gross yes but still practical and useful. Insects have also gained the attention of scientists and have the potential to be used as drug sources and other pharmaceutical substances (Heming, 2018).

Insects are also used as educational tools. Understanding various animal species is part of the curriculum for primary, secondary and tertiary education. Studying insects provides insights that are important in understanding the world of biology. Insects can be studied to observe and understand biological processes such as growth and reproduction which can be observed. Insects are ideal subjects for both classroom and laboratory study and this is due to their life cycle which is relatively short and their physical transformation which is dramatic and differs among species. Other than that they are also readily available in the environment, inexpensive, have ease in handling and maintenance. They are also important for education that takes place in outdoors and it provides students with the opportunity to understand things like food chains or food webs, pollination and relationships between prey and predators.in addition to the field guides, there is also material that is available especially at the primary or secondary level for the same. Dissection of animals including insects is also used to study physiology and morphology. Fruit flies for example are subjects used for research in the study of genetics.

Other beneficial uses of insects include its use by fishermen as bait while fishing. The ideology to use insects as food has been a practice since time immemorial and is a solution that has been proposed to meet the growing demand for food with population growth but is yet to gain popularity and acceptance for countries in the worst (Heming, 2018).

9.4 ROLE OF INSECTS IN RESEARCH

Insects have gone down in history as useful in the field of scientific research. In genetics Mendel, a famous scientist used fruit fly with the scientific name *Drosophila melanogaster* was a model organism used in genetic studies due to its short generation time, ease in their rearing, high fecundity and small size. It has a lifespan as short as ten days. Cytology and genetics studies have been explored with the use of fruit flies as their foundation. This insect species used in the study of higher eukaryotes in different areas including gene interaction, genetic linkage, chromosomal genetics behavior, development and evolution. In eukaryotes, genetic systems can be seen to be well reserved and the basic understanding of these cellular processes in fruit flies were used in the understanding of similar processes in others. Its genome was sequenced in the year 2000 and it had a 70% similarity to the human genome hence significant and of support to the evolution theory (Morris, 2006).

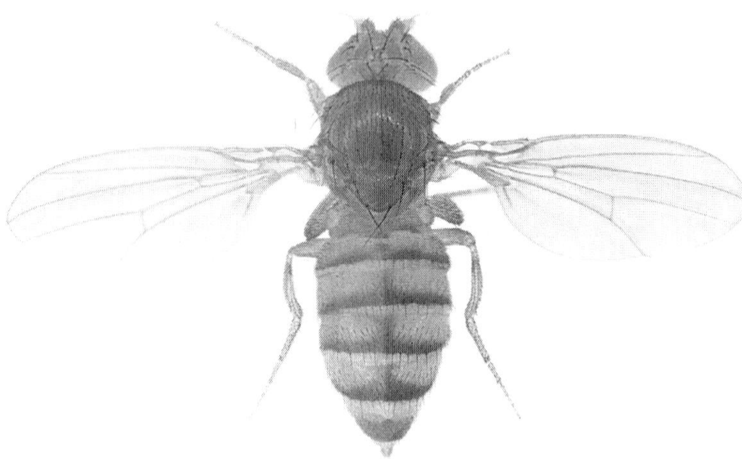

Figure 9.12: *Drosophila melanogaster* used in genetic studies (fruit fly).

[Source:https://ucsdnews.ucsd.edu/pressrelease/new_genetic_method_promises_to_advance_gene_research_and_control_insect_pest]

Insects are also used by researchers as indicators to measure disturbances in the environment, the effects of fragmentation of habitats, and biodiversity changes. The response of sensitive insects to disturbances in their habitat can be rapid, predictable and it can be analyzed and measured easily. Sensitivity to water temperature, chemicals and turbidity are observed in relation to aquatic beetles, larvae to mayflies, stoneflies, dragonflies and damselflies. The absence of species in an ecosystem can givean indication that it is polluted. It may be a result of chemical dumping that is illegal, runoffs from agricultural inputs and mining; sewage treatment, erosion and effluent from power plants (Morris, 2006).

Monitoring of aquatic insects can indicate there is a problem before pollutants manifest themselves in populations of plants and animals. Changes in the population of terrestrial insects are used to measure any changes in plant communities affected by development in urban areas or agriculture. Insects present in soil have also been used to indicate the fertility of soil and pollution levels. Ecological habitats that need protection can be identified using insects such as butterflies or beetles.

9.5 ROLE OF INSECTS AS FOOD

Figure 9.13: A delicacy of flying termites.

[Source: https://www.thepatriot.co.zw/old_posts/nutritious-insects-among-the-batonga/]

Food habits and culture differ within different societies and for some cultures, insects can be considered as delicacies while in other cultures they are just a day to day diet. In human nutrition, insects can provide a highly nutritious meal for animal protein. The consumption of insects is referred to as entomophagy and as much as it is normal in some cultures people from the west consider it a taboo. The task of eliminating pests from agricultural produce sometimes is difficult and they find a way into our food a perfect example would be weevils that attack grains or green caterpillars on green vegetables. In some countries, food laws and standards do not prohibit the presence of insects in food but rather impose a limit on quantity. According to one anthropologist, it is evident that most cultures that do not accept the consumption of insects have a variety of other animal protein sources. Such as seafood, poultry and livestock. Considering the continuous growth in the world's population and the constant occurrence of food shortages people may begin to consider insects as a possible source of food in the near future. From entomology studies, we conclude that insects have high and significant nutrient content e.g. nutrients such as proteins, minerals, fats which are consumed by part of the world population. Insects can be consumed by humans and can be similarly used as a whole or as any other ingredients in foods that are processed for example in burger patties, pasta or as snacks (Berenbaum, 2010).

There are different species of insects that are being consumed worldwide that is between one thousand to two thousand species. Some of the species include about two hundred and thirty-five species of butterflies, three hundred and forty-four beetles, three hundred and thirteen ants, wasps and bees, thirty-nine termites and twenty dragons flies as well as cicadas which are insect species eaten. Their consumption in terms of regions will vary due to different factors such as the environment, ecosystems and also the climate.

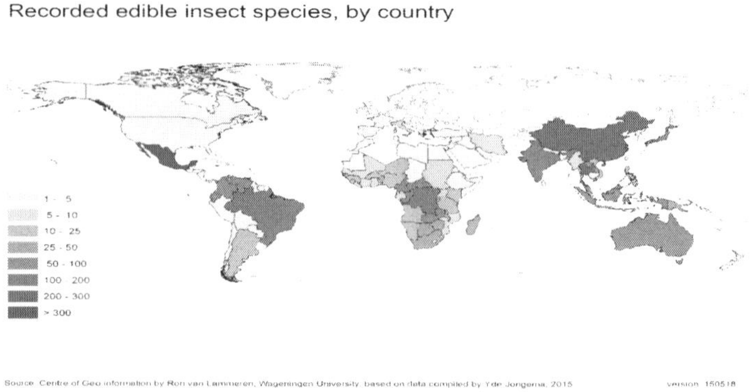

Figure 9.14: A record of the geographical distribution of insect consumption.

[Source: https://www.sciencedirect.com/science/article/pii/B9780128028568000028]

9.5.1 History of Insect Consumption

Before civilization, insects were an important part of the human diet. Historic evidence suggests that the prehistoric man was entomophagous and were therefore insectivorous. Paintings in a cave found in northern Spain which is dated between thirty thousand and nine thousand BC depict that early man collected edible insects and wild bee nests suggesting they were insectivorous. In China, there were cocoons of wild silkworms found in ruins in the province of Shanxi dated between 2000 and 2500 BC. The cocoons were found with large holes in them suggesting that they had been eaten. Practices in entomophagy have changed since then in comparison to other agricultural practices and this has led to the development of insect rearing practices that are modern.

9.5.2 Traditional or Cultural Consumption of Insects

Many ethnic groups in Asia have been consuming insects for the longest time so have some communities in Africa, Mexico and South America. Up to 2086 species of insects have been or are being eaten by 3,071 ethnic groups in 130 countries. In Columbia and North East Brazil, the leafcutter is traditionally and commonly consumed. There is widespread consumption of caterpillars from the *Gonimbrasia belina* and the mopane in the southern part of Africa as a source of protein. Indigenous populations in Australia consume the witchetty grub. The Nunamuit people traditionally consumed the *Hypoderma tarandi* grub and the reindeer parasite. In northeastern India, *Udonga montana* is a bug traditionally consumed in this region but has periodic population outbreaks.

In Indonesia, traditional ethnic groups have been known to consume insects some of which include grasshoppers, termites, crickets, bees and the larvae of the sago palm weevil. The grasshoppers and crickets are usually deep- fried in palm oil after being buttered and it is eaten as kripik or rempeyek snack. There is also special insect meal prepared in East Java called botok, which is beehives containing the larvae of bees and it is also seasoned with spices and coconut that has been shredded. Afterward it is wrapped in a package made of banana leaves and it is steamed. Another popular dish in eastern Indonesia is the ulat sagu or the palm weevil larvae and are either roasted or eaten raw.

Figure 9.15: Indonesian Botok, wrapped in a package made of banana leaves.

[Source: https://en.wikipedia.org/wiki/Entomophagy]

In the northern provinces of Thailand, certain insects are also consumed. Insects are sold in their traditional markets and they include grasshoppers that are deep- fried, crickets, and the larvae of bees, silkworm, ant eggs and termites. In Mexico, insects are used as food ingredients. In east Afric,a kunga cake is made from flies that are compressed densely.

Figure 9.16: Insects sold in stalls in the streets of Bangkok in Thailand.

[Source: https://en.wikipedia.org/wiki/Entomophagy]

9.5.3 Western Culture

In the western region, products such as honey and carmine are consumed, the consumption of insects is not popular in the west but there is a trend towards them. By 2011, a few restaurants in the west served insects on a regular. For example, in Vancouver, British Colombia, Canada hotels offer meals items that are cricket based. Vijs restaurant serves parathas made from crickets that have been roasted and grounded into a powder. Rongoli restaurant has a menu item with whole crickets as a pizza topping. Aspire food group in North America, was the first company to engage in insect farming and developed an entomophagy company. It has machinery that is automated in a warehouse where crickets are grown for consumption by humans. Grasshoppers are commonly sold at Safeco field the home stadium of the Seattle Mariners (Marshall Cavendish, 2000).

Culturally the consumption of insects is seen as a taboo. There exist some exceptions such as the casu marzu which is an Italian cheese that contains insect larvae and in the Sardinian language it is commonly referred to as maggot cheese. In Islam, locusts are considered halal but other crawling animals are considered haram. With Judaism, most insects are prohibited or considered kosher except a few species of locust which are considered acceptable in some Jewish communities. According to Alan Danguor, a public health nutritionist there are still many barriers to insect consumption in the west and one of the most recurring reasons is acceptability as most people do not consider them as clean. In many cultures that have other protein sources such as meat from cattle or poultry, the consumption of insects is sometimes considered taboo. Some cultures have adopted both livestock and insect farming. Some examples include Botswana, South Africa and Zimbabwe where the raising of cattle is paired with the entomophagy of insects like the mopane worm. In countries where entomophagy is a common practice, the choice of insects is not indiscriminate and insects that are not culturally acceptable when consumed are seen as taboo (Marshall Cavendish, 2000).

Figure 9.17: Edible fried insects being sold in the Germany streets.

[Source: https://en.wikipedia.org/wiki/Insects_as_food]

9.5.4 Edible Insects for Industrialized Mass Production

Markets in the west for example in Europe or North America breed insects for academic purposes and also insect production in large scale especially by companies that produce food such as the entoforms in Canada, in the united states there is the aspire food group, the proteiform and protix in the Netherlands and Switzerland the Buhler group. All the above- mentioned companies focus on majorly four species of insects that are okay to be consumed by humans as well as being included in mass production in industries. The insects include the house cricket, the European migration locust, mealworms as locusts and lesser mealworms which are commonly referred to as buffalo worms.

9.5.5 Nutritive Value of Insects

In comparison to meat sources, insect have more efficient nutrients. Crickets, for example, are a complete protein source and these can be found in amounts that are adequate when compared to plant proteins e.g. from soybeans but studies have shown that it is less than that from casein which is a protein that is mostly found in dairy products. The crickets also contain dietary fiber which plays an important role in our diets.it also contains dietary fat especially the unsaturated kind and also some vitamins such as B12, riboflavin and vitamin A together with vitamins that are essential (Shields, 2017).

Another example is locusts which contain iron of about 8-20 mg for every 100 grams of locusts in their raw form and in comparison beef contains approximately 6mg of iron in 100 grams of meat. Crickets have 12.9 grams of proteins in every 100 grams, one hundred and twenty-one calories and five point five grams of fat this is half of what is contained in beef (Shields, 2017).

Food Item	Protein (g)	Fat (g)	Carbs	Calcium (mg)	Iron (mg)
Giant Water Beetle	19.8	8.3	2.1	43.5	13.6
Red Ant	13.9	3.5	47.8	47.8	5.7
Silk Worm Pupae	9.6	5.6	2.3	41.7	1.8
Dung Beetle	17.2	4.3	.2	30.9	7.7
Cricket	12.9	5.5	75.8	75.8	9.5
Sm. Grasshopper	14.3	3.3	27.5	27.5	3.0
Lg. Grasshopper	20.6	6.1	35.2	35.2	5.0
June Beetle	1 3.4	2.9	22.6	22.6	6.0
Caterpillar	6.7	n/a	n/a	n/a	13.1
Termite	14.2	n/a	n/a	n/a	35.5
Weevil	6.7	n/a	n/a	n/a	13.1
Beef (Lean Ground)	27.4	n/a	n/a	n/a	3.5
Fish (Broiled Cod)	28.5	n/a	n/a	n/a	1.0

Figure 9.18: Nutrition profile of different insects in 100 grams.

[Source: forestandrange.org/modules/behavior/palatability/palatplayprint.ht]

9.5.6 Farming, Production and Processing of Insects

Insects to be used for consumption can be raised the same way as livestock in farms that have been specialized in many western countries such as countries in Europe. Insects to be consumed by humans like other animals are produced under food laws and hygiene standards that are stringent. There are different factors and variables that are considered in breeding and production of insects they include temperature, humidity, feed, water, housing and this is dependent upon the type of species. In most cases, insects are first bred from eggs for larvae such as meal warms or are left to mature such as locusts or crickets and afterward they are killed using temperature reduction. After being dried by freezing, the insects can afterward be packed as whole or as insect powder and then it can be used in the processing of other foods such as baked products or snacks which are then sold for consumption (Khaliq et al, 2014).

Figure 9.19: Crickets being bred from human consumption.

[Source: https://commons.wikimedia.org/wiki/File:Bolikhamxay_Thabok_Crickets.JPG]

Other factors considered in the selection of insects for breeding other than the nutritional value of the insects or the easy of digestion, insects are selected by the producer based on how easy it is to rear them this includes their disease susceptibility, the ease and efficiency in their conversion to food, their rate of growth and development and the generational turnover. Some of the food products from insects include insect flour from pulverized freeze -dried insects, insect burgers made from patties which contain patties made from insect flour from worms or crickets; insect protein bars which contain insect flour; insect pasta which is made from flour fortified with insect flour; insect bread which is baked used flour from insects but mostly from crickets; snacks from insects which include crisps or flaps from insect powder together with other insects (Khaliq et al, 2014).

Figure 9.20: Products from the processing of insects from the right a snack from cricket flour, energy bars and insect bites made from cricket flour.

[Source: https://en.wikipedia.org/wiki/Insects_as_food]

9.5.7 Advantages of Consuming Insects

In 2015 Chris Derudder, a Belgian entrepreneur introduced the world edible insect day to create a global awareness for edible insect consumption and is held on the twenty-third of October. This agenda was started with a focus mainly in Europe, the northern part of America and also Australia.

The advantages of eating insects include increased food security which is a result of increased biodiversity giving people a variety of options and it also provides alternative protein sources. It can, therefore, be viewed as one of the solutions to food insecurity given the growing population globally. It lowers resource use and also contributes to sustainability in relation to environmental resources and the economy since relatively fewer resources are required in insect production compared to agricultural or livestock farming. Insects can also be used to develop therapeutic foods and there is still a lot of research being done and funded to prove this for example in 2009, a doctor from Ghent University proposed the use of proteins from insects in the generation of hydrolysates which have an ACE inhibitory effect and also produces antioxidant activity which can also be incorporated with other functional foods and they also provide unsaturated fats which can reduce cardiovascular diseases (Gilbert, 2011).

Cultivation of insects by edible indigenous populations can be of benefit to them economically, nutritionally and also ecologically. Environmental benefits include less resource use and also prevention in the strain of the ecosystem in the production of insects in comparison to other animals. Less environmental degradation and production of greenhouse gases have also associated with breeding and consumption of insects. Economically insects are associated with a higher food conversion efficiency in comparison to traditional meats. With the high production rate of insects, there is also the prospect of financial or economic gains once more people embrace insect consumption. Insect production is also associated with efficient land use and also less water pollution in comparison to conventional livestock agriculture (Gilbert, 2011).

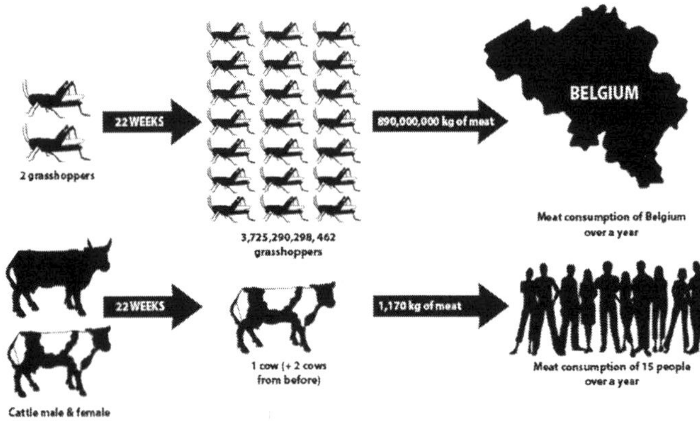

Figure 9.21: Difference in production between cattle and grasshoppers.

[Source: http://payload226.cargocollective.com/1/14/457568/6829789/Food-production-comparison.png]

9.5.8 Disadvantages and Concerns related to Consumption of Insects

There are a few disadvantages and concerns that are related to the consumption of insects but most of them have to do with food safety concerns similar to any food that is consumed such as bacterial spoilage, producing allergic reactions and toxicity. Other concerns are ethical and there also other concerns in relation to the environment such as sustainability.

Food safety concerns: Challenges and safety concerns related to insect consumption-despite insects being beneficial as a source of food, they are potential challenges that are brought by a concern in production and also safety in the same. In the insect industry, mass production is a concern due to lack of technology for harvesting and production of insects in a manner that is efficient since the machinery would have to be based on the life cycle of the select insects as well as a have temperature control which is a key aspect in the development of insects.

Another consideration that has to be made is the insect shelf life in comparison to other meat products which brings about the concern of food safety. Insects have the potential of leading to hazards such as contaminants, pathogens, heavy metal concentration, causing allergies and also pesticides.

For example, beetles can cause chemical hazards such as oromones, cyanogentic substances and contamination by heavy metals; silkworms produce chemicals that can cause allergies the substance is called thiaminase; the honeycomb moth also produces cyanogentic substances; the house cricket has the potential to cause microbial hazards and ants also produce anti-nutritional factors such as tannins.

Food safety hazards can be controlled for example in the case of allergic reactions food labeling of processed packaging is necessary to prevent them from being consumed by people with allergies. To minimize chemical hazards producers can take part in selective farming of insects. Cooking processes can be used to minimize potential microbial and parasitic hazards.

Authorizations and regulations: In some countries regulations have been established in relation to insect production for example Switzerland has approved the breeding of some insects for consumptions such as the house cricket, European locusts and mealworms which are insect larvae. The above insects can also be prepared for human consumption in their whole form, pulverized form or in food products that have been processed. In the European Union, the European Commission defines insects as novel foods which are food ingredients isolated from food. Insect parts also fall under the same definition. The European food safety authority is still investigating insect species and in 2018 it published the risks of using the house cricket for food.

9.6 INSECTS AS FODDER

As the world's population increases, the need to diversify foods becomes a concern and in turn, preferences change especially in the case of animal sources. In order to fulfill this need, we consider improving animal breeds which also need to be feed. Every year a significant amount of feeds that are bought and are expensive are used to feed these animals. Poultry have the most significant percentage as they are fed with grains, soy and fishmeal.

Pigs are usually fed on things like kitchen refuse, grains, soy and also fish meal. Cattle can be fed on fodder from Napier grass, some grains and also soy. In aquaculture, the cultivation of fish, and are fed on fishmeal, grain, legumes, soy and pellet form. Fish, pigs and chicken which are farm animals sometimes feed on insects and maggots which are used as fodder. Why consider insects as animal feeds? Why are insects better than animal feed? Insects are very nutritious as discussed previously in the article it consists of between 40-60% proteins, 36% fats (Heming, 2018).

These insects can be knowingly and unknowingly being fed on by some animals and these can further the discussion of using these insects naturally found in the environment as food for these animals. Other insects commonly used as fodder include black soldier flies, the common housefly and mealworms. Both house flies and black soldier flies are commonly found in organic piles and their cultivation can help in the promotion of improving organic waste management. They can also be a good addition as a feed ingredient to pets or livestock. In the E.U there was a common practice of feeding cows with dead animal meat and processed meat which saw the death of many animals which led to the banning of both animal and insect feeding to animals an exception was fish meat. In 2017, the European Union permitted the use of some insect species for aquaculture and they include the housefly, the black soldier fly, mealworms, house crickets, lesser mealworm and the tropical house cricket.

Other than insects for feed being selected for their composition in terms of nutrition, they are also selected based on their digestibility and also based on how easy they are to rare (Heming, 2018).

The substitution of fodder with insects may be beneficial. When practicing aquaculture, we use wild fish to make fish meal and therefore using insects can reduce overfishing. Insects also do not require a lot of space and energy compared to soy cultivation. A study has shown that the production of 1 metric ton of crickets is equivalent to 600kg of protein and for soy, one metric ton is equivalent 0f 60 kg protein. This means rearing crickets is more feasible than planting soy. Another appealing aspect is that while cultivating insects they can feed on organic waste. Insects have a wide range of substrate sources that helps them to thrive. Insects can also be bred all year round providing a nutritious meal to animals all throughout the year. The use of insects as food can be of economic significance to different countries by saving money that would have been used to import raw material to be used in processing fodder. The use of insects as fodder can contribute to environmental sustainability if it adopted as breeding of insects lesser resources in comparison to other agricultural practices (Gilbert, 2011).

9.7 ROLE OF INSECTS IN OTHER PRODUCTS

Chitin, grease, protein can be provided by insect larvae in the pharmaceutical industry this grease is used for manufacturing cosmetics and surfactants for shower gels, therefore, replacing vegetable oils such as palm oil.

Different species of insects over the centuries have been used to produce products such as varnishes, dyes and inks. The lac insect produces shellac which is a sticky resinous material and it is produced as a coating that is shell-like. The insects live in small groups and encrust twigs of acacia. The insects are scraped off from the twigs biannually to produce bright red lac dye. The shellac is also produced and it is used in furniture, shoe polishes, printing inks, electric insulators and shoe polishes. It is also used in sealants, cloth buttons and hair care products because of its elasticity, solvent resistance, adhesive properties and qualities of insulation. Shellac based records were used for the development of compact discs and vinyl albums.

Dried and pulverized bodies of the cactus meal bug or the cochineal insect are used to produce the red cochineal dye. The cochineal insect is mainly found in the southwestern part of the United States and it feeds on cactus exclusively. The cochineal insect is cultivated by the Pima Indians of Arizona and the Aztecs of Mexico and was used for dye. The use of the cochineal insect for dye was first reported by a westerner Hernando Cortez and the vermilion dye became an important item for trade in Europe. This then spread to other countries throughout the Mediterranean region and also in India or Australia and even the Canary Islands. The importance of this dye has declined over the years with the increased use of other dyes that are synthetic although in Latin America it is still used as a natural food color and for coloring soft drinks. It is also used in cosmetics but it is considered non-kosher by the orthodox Jews.

9.8 ROLE OF INSECTS AS PETS

As much as insects seem like unlikely pets, some individuals have kept a variety of insects as pets. The most common insect pet in the United States is the ants in ant farms although other insects are kept. I do get the appeal as ants have come across as one of the most organized animals similar to bees and termites. Although a few people do not desire to keep insects as pets some societies like the Japanese have been known to keep a variety of insects rather than the common house pets such as dogs and cats. Keeping of insects as house pets can sometimes be of value for example the night singing crickets can be used as a detection of the presence of intruders as they have a tendency to stop singing when interrupted by noise. People from the east and the Mediterranean region have kept katydids and crickets as pets over time. The sale of crickets is a common practice among the Japanese. There is another benefit of keeping insects as pets and that includes the fact that they are easy to find and also do not consume large amounts of food and they also occupy small spaces. The only disadvantage is the inability to pet animals and the fact they cannot provide companionship since they are not as intelligent as cats and dogs. Some insect pets that are can be practically kept by humans include:

9.8.1 Caterpillars and Butterflies

The observation of the caterpillar's transformation to the adult stage in their life cycle can be amazing and be fascinating to humans and also provide lessons on nature at its finest and it comes as a highly recommended exercise to groups or individuals in our younger years. When considering the possibility of keeping caterpillars as house pets one can note the type of vegetation caterpillars attach themselves to in their natural environment to maintain similarities like their ecosystems such as the plant foliage for their feeds to ensure a constant supply of food. Nowadays people keep butterflies for recreational and commercial purposes since they have a beautiful aesthetic and can be preserved and their bodies used as pieces of art.

9.8.2 Field Crickets

They are insects that are unable to fly as they lack wing covers. As "artists" they are graceful pests and with the new era of modern technology, scientists have been able to mimic these sounds and adopt them in making technology that helps people sleep with the presence of natural sounds. With their inability to fly they can be kept in bowls rather than an enclosed glass like the butterflies. This container can be filled with a small amount of soil that a person can cover with a piece of cheesecloth or a mesh to prevent their escape. Crickets feed on fruit, lettuce and sometimes bread. Another possible and reasonable food for crickets can be dog foods.

9.8.3 Ant – Lions as Pets

Doodlebugs are the most familiar ant types to humans, are known as ant lion in a more mature sense. They are able to create cone- like pits in sand and they trap other ants within the pit and feed on them and they also have the ability to survive in between meals in the event they are unable to trap their prey.it takes a period of 2-3 years before they become fully matured ant lions. Therefore, they should be kept in boxes that are shallow filled with sand allowing them to trap other prey for food.

Figure 9.22: A formicarium, which is a housing for an ant colony.

[Source: https://en.wikipedia.org/wiki/Ant-keeping]

9.8.4 Mealworms

Are immature larvae of the black beetle and therefore the term worm can be used loosely when describing them. They can be reared or cultivated for commercial purposes rather than being kept as pets as they can be used as foods for other animals such as snakes, birds or other reptiles.at the same time mealworms can often be classified as pests since they destroy some of the agricultural produce in storage including grains flour and other cereals. to successful rear mealworms one should have an enclosed box that contains food especially cereals half filled. Some pieces of fruits such as apples can also be included in crushed form although these meal worms can still be bought in pet stores. For a full cycle, mealworms take approximately six months.

9.8.5 Praying Mantis

It or they can be kept in a similar space as that of fish or the use of a home-made cage. Space should be filled with soil at the bottom at the same time including branches and small plant material for it to climb on. They are cannibals and should be kept in single spaces or compartments. During the spring period is when egg cases of the mantis can be best gathered and brought back to the house where they start hatching. Other insects that can be kept as pets include the stick insect.

Figure 9.23: The stick insects being kept as pests.

[Source: coolpetbugs.com/stick-insects-as-pets/]

9.9 INSECTS IN CULTURE

Culture by definition is the way of life of a society that is their behaviors, beliefs and customs. Culture is transferred to the next generations through social learning and includes their expression through forms like music, art culture, dance, religion, technologies and dressing. This part of the article describes the role insects play in the culture of different human societies. The role of insects in culture can be scientifically be referred to as cultural entomology and it explores the expression of insects in our culture as humans. Ancient cultures have made use of insects all around the world, they have been used to symbolize gods and also for the portrayal of good and evil events. Stamps, coins, and stamps have insects imprinted on them. Artistic professionals such as writers, poets and musicians engage in the liberal use of insects as an inspiration for their work in stories, songs and poem verses. Artists or painters use insects as models while painting, sculpting, making jewelry, furniture and toys.

9.9.1 Insects in Mythology, Religion and Folklore

Various tribes globally have used insects as totems or symbols in their explanations to do with myths of creation. A myth on creation from a tribe in the south of America uses a beetle in its explanation of the origin of the world. The myth claims that the world was created by a beetle and the remaining grains on the earth were used to create human beings. In New Zealand, they are weevils with a grotesque shape that bear a resemblance to their canoe shapes. The Maori tribe adopted the weevil tuwhaipapa as the god of their canoes that thy have newly made. In the Middle East, insect symbolism became popular and widespread during ancient times for example in Egypt around 3100 BC, hornets were symbolic insects when the first Egyptian kingdom was established (Heming, 2018). The reason behind it is their nature which is considered as fierce or threatening. Insects were also important to the religion of Egyptians such as the role played by the scarab. The butterfly is also another important insect in ancient Greece and it was used metaphorically in reference to the soul. In their plays and fables, the Greeks also used beetles, fleas and flies such as Aesop's well-known accounting of ants and also the grasshoppers. Industriousness or insignificance was also symbolized using insects by the Greeks. In relation to early Christianity, some insects were used to symbolize wickedness and foulness and contrary to these insects in relation to religions in the east emphasized unity spiritually with nature considered insects to also signify good luck. Images of Bees were depicted on the shields of medieval knights. For many European families, insects such as grasshoppers, beetles and moths have also found their way to their coat of arms. Nowadays the same applies to modern groups such as sport teams, college and school logos and also mascots of sports teams for both school- based and professional (Heming, 2018).

9.9.2 Insects in Literature, Language, Music and Art

6000 years ago scenes of honey first appeared in caves of the Spanish. Later in the tombs of Egyptians, there were scenes of locusts being depicted in their banquet scenes. In ancient Greece, insect images also appeared on coins and jewelry. For Egyptians between 1600 and 1100 BC, insect jewelry and other sacred scarab ornaments were worn as amulets, necklaces, headdresses and pendants. The Japanese have also pendants carved in a variety of insect forms. In the middle ages insects were also used for decorative purposes on boxes, bowls and sets for writing (Heming, 2018).

Religiously, incorporation of illuminations of insects to surround the borders of important religious manuscripts by medieval monks. Geographic explorations and development of knowledge that is scientific have also led to numerous publications of the discovery of new insect species one of the most common literature references to insects is Franz Kafka's metamorphosis in which the character transforms into a beetle-like insect.

Today jewelry is still made in the form of insects especially butterflies but others include dragonflies, beetles and cicadas which are carved from stone, glass that is fashioned and sometimes metal. Beetles have bodies that are durable and are sometimes used as pendants, earrings. Butterfly wing art is also sold to tourists in tropical countries. Countries also use insects as art for their stamps in post offices. Insects have also been an inspiration behind the names of music groups such as rock bands and also some companies for example motor vehicles. They are also the basis for some derogatory words such as lousy, nitpicker, beetle-browed and grubby. Insects have also been depicted in movies especially animation some of which have teachable meanings such as preserving nature while others have also been used to inspire fear, loathing and even characters that show sympathy (Marshall Cavendish, 2000).

10

Declines in Insect Abundance

CONTENTS

10.1 INTRODUCTION

There are many different studies that have been able to report a great decline in the populations of insects. The most common form of decline is involving reductions in abundance, however, there are some situations where the entire species becomes extinct. There is a lack of uniformity in these declines. In some areas, some reports have been able to prove an increase in the general population of the insects, and some other insect types seem to show an increase in abundance all over the world.

A meta-study that was done in the year 2020 was able to show that there is a decline in abundance of terrestrial insects, at a rate of approximately 10% per decade, while freshwater insect abundance has shown to increase. It is such that those who wrote this report were able to look at that figure of 10% may not be an accurate reflection of the decline in the different parts of the world. In comparison to Europe, there is much less historical data on the decline of insect population when it comes to areas such as Africa or Asia.

Figure 10.1: There are many important insects facing extinction. In fact, many journalists have referred to it as the insect apocalypse. This is sad considering insects account for a majority of our biodiversity, thus being very important to the world in general.

[Source: https://www.nytimes.com/2018/11/27/magazine/insect-apocalypse.amp.html*]*

Examples of insects that are affected the most when it comes to this decline are dragonflies, beetles, moths, butterflies, bees and damselflies. Some of the likely causes of this decline have proven to be the destruction of the habitat, for example, industrialization, urbanization, the use of pesticides (especially insecticides), and intensive agriculture; climate change; and introduced species. It is important to note that not all of the orders in insects are affected similarly; many different groups are undergoing some limited research, and there are hardly any available comparative figures from previous decades (Habel, Samways, & Schmitt, 2019).

Overgeneralization from limited findings and errors in data analysis could also be due to a decline in the scientific field of entomology, which may lead to an exaggeration of the data that is provided on the decline in insect populations. There have been many different insect -related conservation measures that have been launched in response to the reported declines in the insect population. For example, the German government worked to initiate a program for the protection of insects in 2018, which was referred to as "the Action Program for Insect Protection", and in 2019, 27 ecologists and entomologists from Britain wrote a petition asking for the insect research facilities in the United Kingdom, which will help in properly investigating the major and actual threats to insect populations.

10.2 HOW INSECT POPULATION HAS BEEN TRENDING

The fossil records that are concerned with insects can be seen from as far as millions of years ago. This shows that many different background levels are still going on, of both extinctions and species that are appearing as of now. On different occasions, the records also tend to show that many insects are still going extinct, and this is being understood be caused by the natural aspects such as meteor impact and volcanic activity (Shaw, 2014).

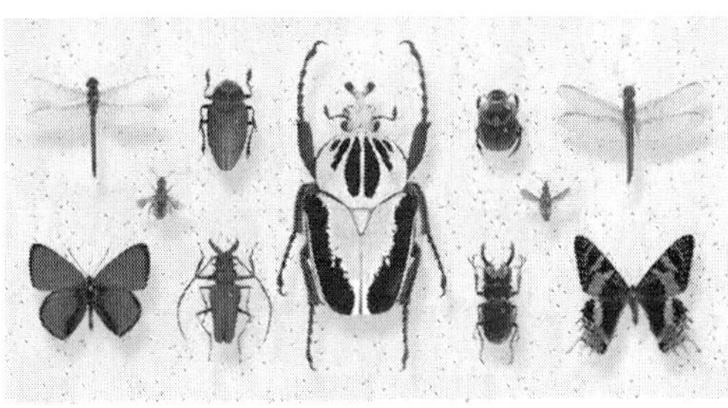

Figure 10.2: It is important to understand the history of insect decline to understand how certain issues are affecting right now; however, there is not much information on insect decline from the past, especially is certain areas such as the tropical and arctic regions, making it harder to understand the decline in these areas.

[Source: http://www.life.illinois.edu/ib/468/*]*

The greatest level of insect extinction was seen in the Permian-Triassic extinction event, and the second -highest was seen with the Cretaceous-Paleogene event. There has been a certain level of recovery in the insect diversity after a mass extinction, due to the periods in which there is the origination of new species in increased frequencies, although recovering these numbers can take millions of years.

Since the late 20th century, there has been a growing concern about a human-created Holocene extinction, despite the fact that the majority of the earlier concerns had a minor focus on the insects. In 2012, the Zoological Society of London was able to use a report on the world's invertebrates to suggest that there is a great decline in insect populations globally, and this is causing a negative effect on the food supply for other animals and even pollination (Rasnitsyn & Quicke, 2007).

There is an estimation that approximately 20% of invertebrate species have the risk of going extinct, and those that were most at risk were those that had the smallest ranges and least mobility. Information and data on the decline of insects have been available for many decades—there was a study that was able to track insect decline within 1840-2013–however, it was the German nature reserves tat was able to republish the study, and this was able to show that there was a high level of media attention in regards to the matter. The press was using alarming headlines to report this decline, such as "Insect Apocalypse". In 2017, The Guardian got a report from Ecologist Dave Goulson, stating that they were on the course for ecological Armageddon since we, as humans, were making vast tracks of land impossible for many different life forms to live on.

In various studies, aspects such as species richness, biomass, and abundance were often found to show a decline in some locations, however not all; it showed that some species are declining, while others can manage the opposite. Some of the commonly studied insects include stoneflies, damselflies, dragonflies, beetles, bees, moths and butterflies. Changes in the environment affect the different species in different ways, and it is not possible to ensure that all the insect groups will have a consistent rate and level of decrease. Changes in the conditions can lead to easy adaptation to change by some species, leaving other species to struggling for survival.

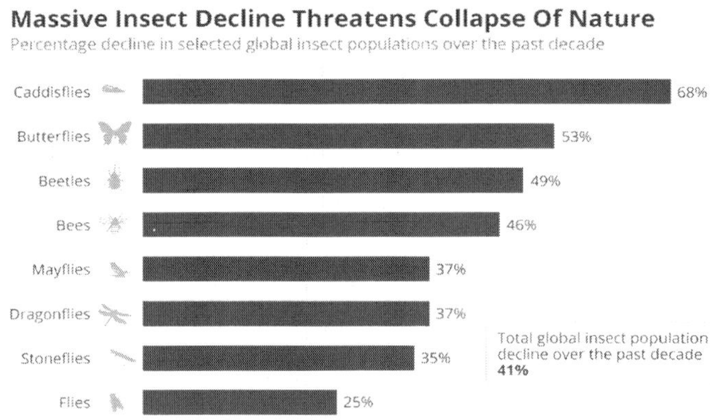

Figure 10.3: Nature is threatened by a massive insect decline, where there has been a 41% insect population decline globally in the past decade. Above are some of the mainly affected insect species. This can greatly affect the insect biodiversity, thus affecting the general biodiversity.

[Source: https://www.pulselive.co.ke/bi/strategy/a-new-report-explains-how-the-massive-decline-of-insects-worldwide-threaten-nature/kkrxvre.amp*]*

The Entomologist Society of America, in March 2019, stated that there was not enough information to use for the prediction of an insect mass extinction, and a number of the predictions that had been extrapolated may have been over- exaggerated or overhyped. European studies have been able to document declines in diversity and abundance of some insect groups such as beetles, bees and butterflies. There are however other different areas that have documentation of an increase in various species, however, there are also other regions that have unknown trends. It has proven to be hard to make assessments of long-term trends in the diversity and abundance of insects, due to the fact that there is not enough data on the historical measurements for many different species. Accurate information for the assessment of species and areas that are at risk is not available for tropical and arctic regions as well as a larger part of the southern hemisphere (Engels, 2012).

10.3 CAUSES OF INSECT DECLINE

10.3.1 Habitat Loss

In the past years, semi-natural and natural habitats have ended up being cleared at a very high and increasing rate to making it possible for the construction of gold courses, lorry parks, factories, housing estates, roads, farming, shopping centers out of town, and many other human endeavors.

Populations of insects that persist on isolated islands, highly fragmented, and even small habitats have a higher chance of going extinct over a certain time period. All in all, the insect population fizzles out, and if they are in locations that are isolated from one another, the chances of them recolonizing is quite low. Recolonization may end up happening years after the initial creation of the islands, and this shows a gradual pace of paying back the debt of extinction. The acceleration of this process may occur if the islands undergo pollution with certain pollutants including agrochemicals from lands surrounding them or even through other forms of degradation (Berenbaum, 2010). There has been a radical change in farming over the past 100 years. In the past, farming practices were less intensive, and this led to the persistence of certain habitats that were favorable to many different insect species, such as bees, and other insects, and intensive farming has led to their decline, these habitats are inclusive of chalk downlands and hay meadows, flowering hedgerows that were able to separate small fields, and even fallow fields that had the ability to flower weeds. In the UK, since 1950, there has been an estimated loss of approximately 150,000 miles of hedgerow, 50% of ancient woodlands, 98% of wildflower meadows, and 50% of downland.

Figure 10.4: Scientists have now been able to confirm that one of the new man- or drivers of reduced insect populations is the loss of habitat, through convert-

ing their habitats to agricultural land with intensive farming techniques. Other additional causes include climate change, invasive species, and agrochemical pollutants.

[Source: https://e360.yale.edu/features/insect_numbers_declining_why_it_matters*]*

In various parts of Britain, agribusinesses have been able to get hold of many different traditional family farms, and these are typified by fields that are often large, and mainly managed by contractors externally, where they undergo maintenance as monocultures that are nearly perfect due to the consistent use of fertilizers and pesticides. This ends up leading to a landscape that is able to produce more food at a lower price than before; however, it becomes less inhabitable to wildlife and fewer people get employment (Berenbaum, 2010).

There is hardly any true reflection of the environmental costs that's been brought about by the production of the cheap foods we are always buying at the supermarket. In addition, we need to understand that farmers only receive a small part of the food retail sale price, and therefore, the expenses of improved on-farm practices would hardly affect the prices that the shoppers receive.

10.3.2 Pesticides

Many different pesticides, such as herbicides, fungicides and insecticides, can easily be purchased at supermarkets, DIY stores and even garden centers. Untrained members in the public end up buying and using them without even taking time to read and understand the instructions that they come with, meaning that they may end up overusing them, or they may also end up using them without wearing personal protective equipment such as gloves.

In intensive farming, the pesticides that are used have shown to ay a huge role in causing the decline of many different insect species; especially because the main purpose of these insects is to kill. The use of pesticides in the United Kingdom has been documented to be much better than in other parts of the world. Even though the tonnage of the use of pesticides has declined in the past 26 years, the more disturbing trend has proven to be the increase in the toxicity of the used products, the amount of treated land, and the number of application (Khaliq et al, 2014).

According to statistics that were provided in 2015 by the Department for Environment, Food and Rural Affairs, there were applications of pesticides of approximately 17.5 per hectare of land that is arable, which means that a total of approximately 17 thousand tons of active ingredients were used.

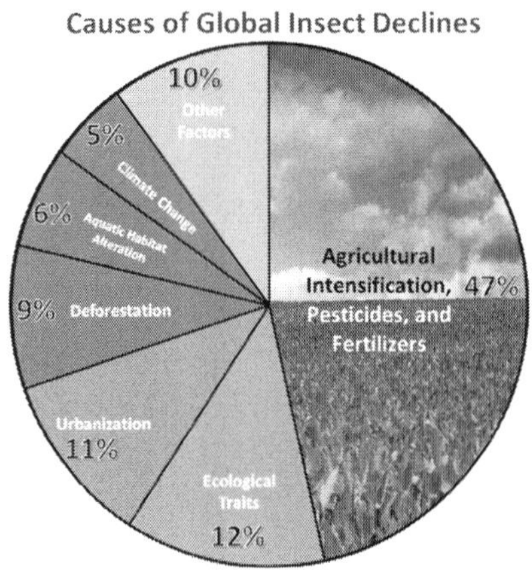

Figure 10.5: The figure above shows that fertilizers, pesticides and agricultural intensification account for 47% of the cause for global insect decline. These chemicals can be washed away and affect different water sources including groundwater, and they can also end up in the rain due to condensation, where they affect the crops. These are just but a few of the ways in which this happens.

[Source: https://hygeia-analytics.com/2019/02/13/intensive-agriculture-is-the-root-cause-of-worldwide-declines-in-insect-populations-new-global-analysis-shows/*]*

The numbers that have been provided above are not inclusive of the pesticides that are used for domestic use, along road verges, streets and parks. There is approximately 18,000 tons of poison that are poured on land every year, it is hard to believe that our animals are not already extinct. When pesticides are used, they end up accumulating in the soil, seep through and leak into watercourses, and may even end up drifting into hedges. Recent research has also proven that even the smallest amounts of pesticides used can end up causing impacts that are highly complex and unpredictable on the behavior of insects (Khaliq et al, 2014).

Majority of the insecticides used globally are neurotoxins, and these can end up leaving the bees that didn't die confused and dazed preventing them from getting back to their hives. In addition to this, there is a high possibility for interactions between different pesticides used together, and even between these pesticides and certain stressors, for example, diseases, and in the presence of two or more stressors, their combined effect may end up being more lethal and increasing the negative effect altogether. For example, the immune systems of bees may be impaired with the use of a tiny amount of neonicotinoids, making them susceptible to certain diseases such as the deformed wing virus. This information is not captured adequately nor accurately by the pesticide regulatory system, which instead ends up focusing on the short term exposure of organisms that are otherwise healthy of single chemicals.

10.3.3 Freshwater Habitats

In the UK, more than 3,900 invertebrate species end up spending a good time of their life cycle in freshwater. This inclusive of many well-known freshwater invertebrates including mites, worms, crayfish, pond skaters, mayflies and dragonflies. These invertebrates are also very important because they play the important role of ensuring that the water remains clean; they help in breaking down and filtering organic matter, in addition to providing a food source to mammals, birds and fish. The health of the habitat that they live in is highly dependent on their presence.

Figure 10.6: Certain orders have some species that only live a part of their life cycle in water, for example, megalopterans, caddisflies, dragonflies, stone-flies, mayflies, etc. while there are others that have both terrestrial and aquatic species, for example, dipterans, orthopterans, neuropterans, butterflies, bugs, beetles, etc.

[Source: https://kids.nationalgeographic.com/explore/nature/habitats/freshwater/*]*

However, in the face of development, abstraction, invasive species, and pollution, there is a decline in many of our freshwater invertebrates. There is often pollution of freshwater habitats that drain from agricultural land, with insecticides, fertilizers, and even metaldehyde that is found in slug pellets. This is eutrophication and it can cause major negative effects to life underwater and life that is dependent on the water. Actually, the close relationship that is seen between eutrophication and insect diversity in streams is so close, that we tend to use insects as bioindicators of aquatic pollution (Habel, Samways, & Schmitt, 2019).

10.3.4 Various Pollutants

In addition to fertilizers and pesticides, there are many other activities done by humans that end up producing many other pollutants, from the approximate 40 million tons of 143,000 various manmade chemicals, all which are manufactured deliberately for many different purposes – the majority of which have led to many negative effects on the environment, to heavy metals such as mercury which end up being released to the environment through industrial and mining processes.

High levels of polybrominated diphenyl ethers and polychlorinated biphenyls (in addition to plastic bags) have been located in crustaceans such as shrimps and crabs that live in the deepest area of the world's oceans: Marianas Trench. For the many different pollutants that exist, there is not much information on how they affect humans, wildlife and invertebrates.

10.3.5 Other Stressors

Insect declines are also being contributed to by many other stressors that hardly receive enough attention as of now. In some ecosystems, there has been a profound reduction in the biodiversity of invasive species; for example, the notorious spread of South American cane toads in Australia—insect species which are voracious predators.

There has been a significant impact due to light pollution on nocturnal insects, for example, moths. The honeybees' cognitive abilities have also been seen to suffer a certain level of impairment from low electromagnetic fields, for example, those that are produced by cables with a high voltage, and there have been suggestions that these may have led to the contribution of losses in bee colonies, and it could also have a major impact on the dispersion and navigation of insects.

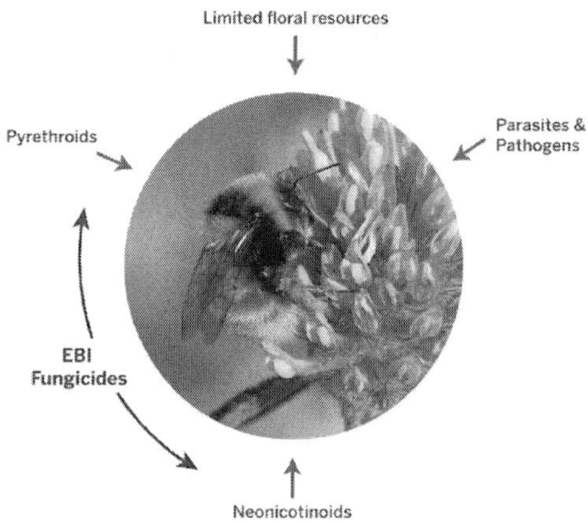

Figure 10.7: Above are some of the stressors causing a decline in honey bee populations. Bees are very important pollinators, and their decline is raising many concerns.

[Source: https://science.sciencemag.org/content/347/6229/1255957?sid=47233960-8655-4e5d-9a4a-3428657a3b54*]*

There is still some research that is being done on other human activities that negatively affect insect health in ways that have not yet been discovered, for the development of newer technologies and also for increasing the speed of development, all which scientists are using to make assessments of how they affect the environment, and their inability to allow for nature to adapt.

10.3.6 Climate Change

Among the worst manmade pollutants are greenhouse gas emissions that lead to global- scale climate change. Until a few years ago, there was no strong evidence on the major impacts that have already occurred to insect populations due to climate change. The analyses that have been made of the insect data in Germany were specifically investigating the matter of whether climate change is the main reason for the declining insect numbers. Even though the day to day patterns of the weather was causing a huge impact on the quantity of insects that were captured, the general climate during the study period hardly changed, and that meant that it was not possible to explain the decline with this aspect (Khaliq et al, 2014).

Studies have shown that the different insect ranges have started to shift while responding to the climate, having the North American and European bumblebees disappearing from their range in the southern edges and occupying high altitudes in regions that have mountains. In addition, evidence has shown that emergence time of various pollinating and herbivorous insects is becoming different from that of their plant hosts; such as, some of the Colorado mountain plants end up flowering before the emergence of the bumblebees from hibernation, and this was not the same in the past. In a similar manner, the leaf -feeding moths have to make sure that their caterpillars emerge at a time that matches the bud burst of their plant host, and this is because it is the time in which the leaves are most palatable. If different cues are being used by the moth and the host plant for timing their emergence — for example, if one is using the length of the day, while the other is using the temperature — it may lead to unsynchronized timing, and this may lead to consequences that are potentially severe for the insects, as well as many other life forms that feed on these insects.

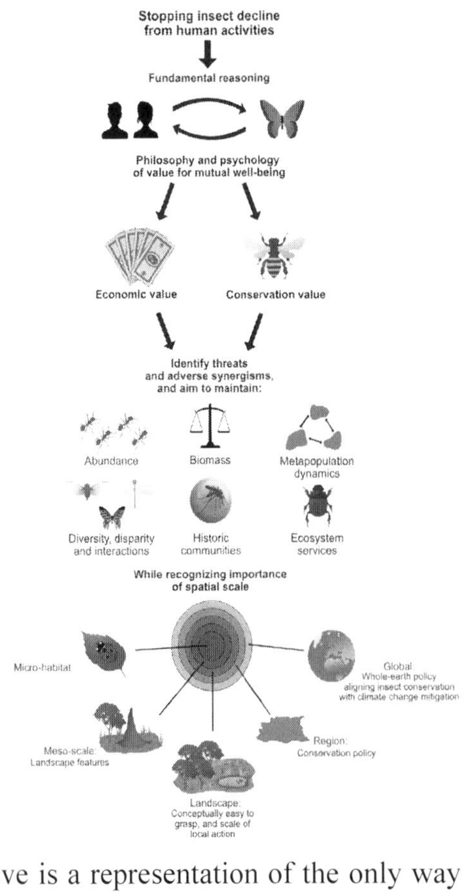

Figure 10.8: Above is a representation of the only way we can end insect decline as a whole. It starts with us reasoning in a fundamental manner. We also need to make identifications of adverse synergisms and threats so that

we can aim to maintain other aspects such as biomass, abundance, ecosystem services, historic communities, among others.

[Source: https://www.sciencedirect.com/science/article/pii/S0006320719317793*]*

There is evidence that this phenomenon may have started to affect the winter moth and oak leaf-roller moth populations. As of now, there are only subtle effects that have been observed, however, they may become much worse with the accelerated climate change that is being seen right now. It was only until a few years ago, that the opinion on the fact that climate change could not account for the majority of the decline in insect populations, changed. It was in 2018 that there was an emergence of substantial evidence that showed a major decline of about 81% and 97% insect biomass for two main samples (sticky trap and sweep-net) in the Puerto Rico rainforests since 1971.

In the last 30 years, there has not been logging or alteration of these forests by humans, and there has also been no use of pesticides on or close to them, in accordance with the knowledge of the people around these areas. This is different from Germany, where there have been changes in the climate of these forests since 1971, with the temperature increasing by 2°C in the mean of the maximum daily temperature, and the researchers who were able to conduct this research were able to conclude that these changes were the main reason why there was a decline in insect populations, however, there could have been other unidentified factors that could have played the role (Habel, Samways, & Schmitt, 2019).

10.4 THE REASON TO CARE ABOUT INSECT DECLINE

Entomologists and ecologists should have a deep concern for the fact that they have not been able to do a good job of explaining the major importance of insects to the global population. Of the known species, insects end up making up a majority of them, and they are also highly important in freshwater and terrestrial food webs. Lacking insects means that fish, small mammals, amphibians, reptiles, bats, and many different birds would disappear because they would not have anything to eat. 88% of the total plant species are dependent on pollination, the majority of which is done by insects. This means that it is pretty much all the plant species apart from conifers and grasses that are dependent on insects' pollination.

There is an approximate three-quarters of all the types of crops that are grown by humans, that depend on insects to pollinate, and this is a service that has undergone estimation to be worth between $234 billion and $578 billion every year globally. When we eliminate the financial aspects, without pollinators it is not possible to feed the global human population. In terms of the ecosystem services that insects provide, their importance is often justified, and this is can end up being ascribed a monetary value. There is actually more than pollination when it comes to the importance of insects; insects such as lacewings, ground beetles, hover flies, and ladybugs have proven to be important agent of biocontrol with their ability to control other insect pests. Wasps and wood-boring beetles have also helped in recycling nutrients that are found in timber that is decaying, while certain armies of invertebrates such as woodlice, worms, silverfish, and springtails, help in breaking down the leaves that end up falling every autumn (Adkins-Regan, 2013).

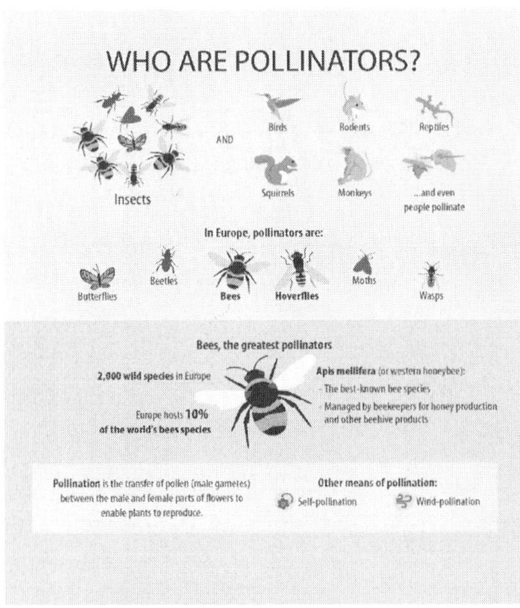

Figure 10.9: There are many different types of pollinators; birds, rodents, monkeys, squirrels, reptiles, and even people. However, the greatest pollinators are insects; bees to be specific account for the majority of the pollinators globally. Others include butterflies, beetles, moths and wasps, all of which are endangered insects.

[Source: https://www.europarl.europa.eu/news/en/headlines/society/20191129STO67758/what-s-behind-the-decline-in-bees-and-other-pollinators-infographic*]*

Were it not for the quick arrival of flies and dung beetles, animal dung would end up building up in our pastures. Flies and dung beetles are able to quickly recycle it and turn it into nutrients that help in the growth of grass. Carrion beetles and maggots rapidly consume animal corpses, which otherwise end up taking months to rot. Burrowing insects such as ants help in aerating the soil and even in the dispersal of seeds. Silk is provided to us by silk moths and bees provide us with honey.

These are services provided to us by the ecosystem and are an estimate worth of at least $58 billion each year in the United States. In other words, it is simple to say that we don't know what most insects do. Approximately 4/5 of the 6 million insect species that are thought exist have not yet been given names, let alone been studied for the roles that they can ecologically perform. We are not even close to comprehending the many different interactions that happen between thousands of living organisms that we find in the majority of the ecological communities, and this means that we are not able to state the insects that we require and those that we do not.

Research that has been done on the pollination of crops has shown that the majority of the pollination that occurs is normally done by a few species, however, the reliability and resilience of that pollination is much better when there are more species present. When we rely on one pollinator, for example, bees, this can be a strategy that is risky due to the fact that if the bees are nowhere to be found, then there would be no other insects to pollinate the crops. With changes in the climate comes changes in pollinator communities, as well as species that may not seem important to us today yet, could be very important pollinators in the future (Shaw, 2014).

We have been able to see very many important economic and practical arguments that vouch for the conservation of insect species that either will be or are, important to the survival of humans and other animals. It is important to understand that despite this argument, this anthropocentric approach to insect conservation ends up lacking all the important arguments that are needed for the conservation of biodiversity. Many scientists have proven that there are very many different insects that could go extinct without us being impacted economically. For example, the St. Helen's giant earwig which went extinct without any of us noticing. In New Zealand, they have giant wetas which are also very close to going extinct, but only a few entomologists would be heartbroken by this phenomenon, the rest of us would be unbothered. In the South Downs, Wart biter crickets could easily also go extinct, in addition to the block bog ants which are found in

Hampshire and Dorset in the wet heaths, all of which would not bring about any negative ecological impacts (Wigglesworth, 2012).

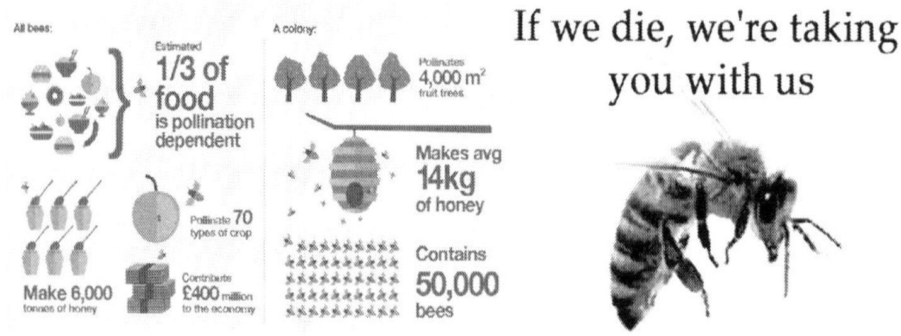

Figure 10.10: The image above shows us how the extinction of bees would affect even us as humans. 1/3rd of the food pollination is highly dependent on bees. They contribute approximately £400 million to the economy. These are just but a few reasons why bees are very important to us and to the environment.

[Source: https://sites.jmu.edu/gbio103/what-would-happen-if-the-bee-population-continued-to-decline-design-to-shock/*]*

One thing we do need to understand ,however, is that as humans we could be able to survive in a world that has very low biodiversity, such as areas like Cambridgeshire or Kansas, which are already living a so. Why not just eradicate entire species at will; for example, the mosquito *Anopheles gambiae* can be fully exterminated through gene drive technology, making it possible for us to one day utilize this technology to wipe out these mosquitoes all over the world. If possible, should we go through with it? As of now, there is even development of robotic bees by robotics engineers for crop pollination, all of which is being done due to the fact that data has shown that bees are in decline and there is a need for us to replace them if we are going to survive.

We need to ask ourselves if this is the future we want to live in, or even for our children, one in which our children will never be able to see a butterfly fly over a hedge, a world without any wildflowers, and where no sounds are coming from birdsongs and even buzzes coming from insects, all which will have been replaced by robot pollinators having a highly monotonous drone (Price, 2003). When we think about these questions, they allow us to argue in a way that values nature for what it provides for us all over the world, either for inspiration and joy or even practically. In conclusion, there is also an argument that does not only focus on human wellbeing: we can bring up the argument on the protection of all living organisms on earth. We have the moral right and duty to protect all living organisms, from silverfish, pandas, penguins, to bees, wasps and ants. All of them have an important role on earth whether we know it or not.

10.5 CRITICALLY AFFECTED INSECT SPECIES

10.5.1 Earwigs

These are one of the smallest insect orders (Dermaptera), with around 12 families and 2000 species, thus being one of the smaller insect orders. They have membranous wings that have been folded under forewings that they rarely use, two pincers that are forceps-like on their abdomens, and cerci that are all quite characteristic. Some specific species act as mammal parasites, therefore lacking the pincers. These insects are found on all land apart from in Antarctica. They are very beneficial due to the fact that they feed on pests that have soft bodies, for example, aphids, and even on the eggs of such insects. It is also possible for them to have economic importance due to the fact that they also feed on soft plant tissues, such as fruits or leaves, which can be very detrimental to home gardeners.

When large numbers of earwigs try to enter households for shade, moisture or shelter. They may also end up feeding and surviving on food items that they find in the pantry, however, no study has proven them to lay eggs and reproduce from inside such areas (indoors).

Figure 10.11: Earwigs can actually have benefits, and are considered very useful insects. They are actually referred to as sanitary engineers in many different parts of the world; this means that they are often helpful in cleaning up the environment as they feed on dead and live insects, as well as on decaying plant material.

[Source: https://sqwabb.wordpress.com/2015/05/16/earwigs-not-lovely-bugs-but-mostly-harmless/amp/*]*

Despite the fact that they have hind end protrusions that are menacing and pincer-like, they have not been known to directly harm pets or humans. These protrusions are referred to as cerci, and they use these structures as a mechanism of defense and also for capturing prey.

10.5.2 Hemiptera

These are also referred to as true bugs, and this order of insects has approximately 60,000 to 90,000 species, and the groups include shield bugs, leafhoppers, planthoppers, aphids, and cicadas. They all have similar sucking mouthparts, however, they range in size from 0.04 in to 6 in. Sometimes, we only use the name true bugs to refer to the sub-order Heteroptera. They have many different importances.

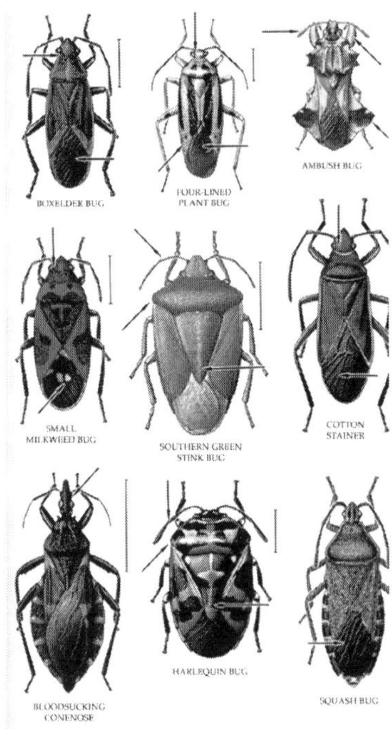

Figure 10.12: Above are species that are found in the Hemiptera order and, and these are referred to as true bugs. They are herbivores and have been known to cause damage to plants, which in turn causes farmers to use insecticides which can affect insect biodiversity by killing them and other insects in the area.

[Source: https://www.pinterest.com/bgphila/hemiptera-true-bugs/]

There are many species that are found in the family Corixidae. The majority of these species feed on algae, and others as predators and even collectors. The remaining semiaquatic and aquatic bugs are mainly predators as adults and also nymphs. They penetrate the prey using their piercing mouthparts, injecting hydrolytic enzymes and sucking out their prey's juices. The majority of them have raptorial forelegs, allowing them to easily handle their prey. Many of them are quite fierce, take large prey, and cause really painful bites. Semiaquatic species tend to feed either from below the surface or from the surface. Purely terrestrial bugs may either be parasites, herbivores, or even predatory.

Many of these insects diverse in lentic or slow lotic waters that are shallow, heavily vegetated and warm. Corixids tend to show much more preferences with their habitats and are also more specious. The majority of them only come for air on the surface.

Terrestrial species have proven to be very important disease vectors, pest -controlling predators, and also as agricultural pests. Aquatic species can help in controlling insects acting as a nuisance, others are utilized as a pet or human food, others end up feeding on hatchery fish, some are a biting nuisance, however, fish only feed on corixids.

10.5.3 Blattodea

This order is inclusive of termites and cockroaches. Initially, termites were classified in the order Isoptera, but molecular and genetic evidence was able to prove that they have a very close relation to cockroaches, both of them having an ancestor that is common. The mantises and the Blattodea are now all classified under the same superorder (Dictyoptera). Approximately 4,500 cockroach species in around 600 genera and around 3,100 termite species in approximately 400 genera are part of Blattodea.

Of all insect pests, cockroaches are among the most cosmopolitan. They have a high association with human households all over the world, where they hide in crevices and cracks in the day and emerge at night to look for water and food. Despite the fact that they do not bite or sting, they have a high association with conditions that are highly unsanitary and may end up carrying many different human and animal pathogens as they move. When cockroaches die, they start to decompose (normally, their feces also decompose), and these can end up causing allergies to people to have a high sensitivity. In addition to all this, it is important to note that they have also proven to be very important tools of research, especially for studying toxicology and physiology.

10.5.4 Orthoptera

This is an insect order that is inclusive of crickets, locusts and grasshoppers, including insects that are closely related such as wetas and katydids. This order can be divided into two main suborders: Ensifera – crickets and their close relatives; and Caelifera – locusts, grasshoppers and their close relatives.

Figure 10.13: Insects in the Orthoptera order are considered the most dominant group in the majority of the habitats that are terrestrial. They normally feed on all plant types, therefore causing a lot of economic damage which can affect the countries economy as a whole. Parts of North America, Asia and Africa are greatly affected by swarms of locusts or grasshoppers, destroying wide land areas of crops.

[Source: https://en.wikipedia.org/wiki/Orthoptera*]*

A number of these species are rangeland and crop pests, or they also seek warmth in human households. The major pest species Orthoptera, causing the most damage are locusts and grasshoppers. In history, locusts have been known to wipe put crop fields in a single day. They are able to eat up to the weight of their own body in a day. Locusts normally gather in swarms (large groups of locusts) and these can have up to 70 million locusts, a swarm which can spread over an area of 470 square miles. Grasshoppers on the other hand can lead to a wide extent of agricultural damage, however, not as bad as that which has been seen with locusts. Both of these insects normally feed on grasses and weeds; however, when there is a high population density and/or drought, they end up feeding on crops. The commonly fed on crops are soybean, and they mainly feed on these once the grasses and weed they normally feed on have become scarce.

This order is the only one inclusive of insects that can be referred to as kosher in Judaism. The book of Leviticus has a list of dietary laws that do not allow for one to eat any flying insects that work, apart from specific locusts. Weirdly enough, the crane fly and dragonfly are not considered kosher, yet they end up being helpless when they are not able to fly. It is also stated by the Torah that the only kosher flying insects have four legs that can be used for walking, with knees, and allow them to hop due to their extension above the feet.

There is a lot of research that is aiming at finding different sources of biofuel in insect guys, and one of the species that has gained a lot of interest is the grasshopper. The grasshopper can break down lignin and cellulose without the production of greenhouse gases, and this has attracted a lot of interest.

10.5.5 Hymenoptera

This order of insects is quite large, including ants, bees, wasps and sawflies. This order has more than 150,000 living species that have been described, all not inclusive of the 2100 extinct species.

For their beeswax and honey, the honeybees have been valued way before the Christian times. In modern times, beekeeping has become an enterprise that is lucrative and highly developed. Honeybee workers produce royal jelly, and this has become popular as a cosmetic, however, scientists have not been able to fully prove its beneficial properties.

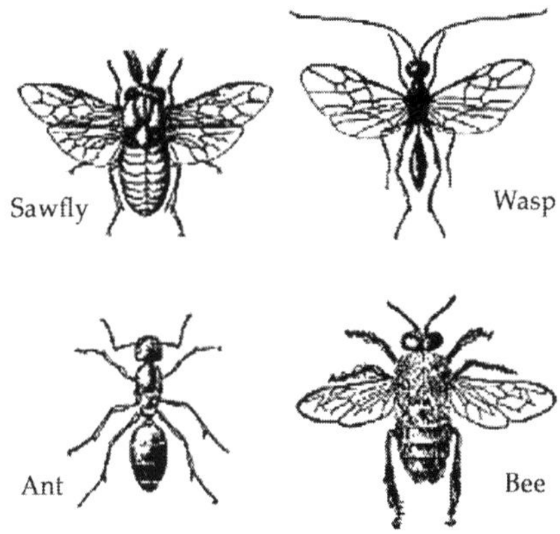

Figure 10.14: Above are the four main species in this order. As stated above, they are known to be the most important pollinators globally. With all the pesticides being used, and climate change which is affecting their ideal temperatures, they are becoming endangered species, and this could affect us all.

[Source: https://projects.ncsu.edu/cals/course/ent425/library/compendium/hymenoptera.html*]*

In this order, some parasitic forms have been known to help in controlling insect pests. Some of these species are tiphiids that parasitize the Japanese beetle; tichogrammatids and chalcids that parasitize many different orchard pests; pteromalids that parasitize a number of crop pests; eulophids that parasitize scale insects; braconids that parasitize many wood- boring beetles and Lepidoptera; and the parasitic wood wasps that attack wood- boring beetles. In the Western United States, Big insects a well-known and very important as the only Smyrna fig plant pollinators.

There are very few of these species that are serious economic pests. The most important among these, however, are The European spruce sawfly, which was historically a major pest in North America; the larch sawfly, which ended up destroying the majority of the larch forests in North America and Britain later on in the 19th century; a few of the seed chalcids that destroy wheat; and the wheat stem sawfly. To be able to control the Europeans spruce sawfly, the Europeans were able to introduce parasitic wasp.

There was the accidental introduction of the fire ants into the United States from South America, and these feed on seeds and young plants in addition to attacking young mammals. In South America, there is particular importance in the destructive habits of army ants and legionary ants. These ants come in armies of as many as 1.6 million, destroying a majority of all my mammalian life they come across. In Brazil, leaf-cutting ants have also proven to be very serious pests, where they destroy a large quantity of plants that have been cultivated, in one night. In South America, these insects are locally used as a food source.

10.5.6 Lepidoptera

This order is inclusive of moths and butterflies. There has been a description of about 180,000 species in 127 families and 47 superfamilies, and they all account for 10% of the total species that have been described within living organisms. Globally, this is considered one of the most widely recognizable and most widespread insect orders in the world. This order has many different variations when it comes to the body structures of the insects involved, all of which have undergone evolution to attain certain distribution and lifestyle advantages. Estimates that have recently been provided are suggesting that this order may be having many other species that we earlier had no idea about, thus being one of the four orders that have many species, together with Coleoptera, Diptera, and Hymenoptera.

Figure 10.15: These are the common butterflies and moths we know. We see them everywhere we go, but with years, their numbers have started to drasti-

cally drop. These are also important pollinators, and their decline can be felt globally. Many of them are also highly valuable due to their beauty. Others for commerce, such as silkworms that provide silk.

[Source: https://www.pinterest.ie/pin/297519119128412801/*]*

So as to fully comprehend the magnitude of the diversity, these are one of the most important groups of insects. As stated before, to comprehend the magnitude of biodiversity, some of the most important groups that need understanding are insects. Of the total biomass in the tropical rainforest, they can represent up to 94%. They also account for 1 million of the known living organism species on earth against a total of approximately 1.7 million known species. Now that this has been broken down, it is easier to understand that approximately 180,000 of these species are from these specific order (Lepidoptera), constituting one out of four mega-diverse orders globally and also representing the largest contribution to primarily herbivorous organisms. They have enormous ecological importance, not only due to the fact that they have many different species and a high biomass percentage that they account for in the ecosystem, but also due to the fact that they are food for insectivores, pollinators, and also herbivores.

When a butterfly species occurs in a certain area, this can help in indicating of that there are also other plant species in the same area (sources of food for the adult and the caterpillar), animals (predators and parasites), and even a certain specific set of environmental factors. Given the fact that they can easily be observed in the field, the relatively well -documented taxonomy, Spatiotemporal stability, diversity, and even their abundance, butterflies are commonly utilized as bioindicator organisms. In 1999, Blair stated that butterflies were great at indicating anthropogenic transformations on landscapes, and even urbanization in general. Butterflies have a high level of sensitivity to luminosity, humidity, microclimate, and temperature, all of which typically change when their habitats are disturbed. In addition, the butterflies' diversity can help in approximating the diversity of the plants around it due to the fact that butterflies are directly dependent on plants.

Amongst all the insect species, butterflies have ended up becoming a model taxon for conservation and biodiversity studies, monitoring of the animal population, environmental impact estimates, and many other genetic and ecological studies.

The beauty and attractiveness of butterflies have been an arousing aspect for many years, interesting both scientists and amateurs, and the accumulated information on their distribution, taxonomy, and biology can only be compared to that of vertebrates. They have proven important for increasing awareness of environmental issues and even for educational purposes. They have helped in increasing awareness, which is much more positive than that which was there before, on the invertebrates to the public, all mainly due to their aesthetic value. Due to the fact that butterflies have become the most popular insects, they have gained the flagship honor, being used in the conservation efforts for invertebrates.

10.5.7 Beetles

Figure 10.16: Dung beetles have proven to play a role that is very important and beneficial to the environment, both agricultural and natural ecosystems. The majority of these beetles use the fecal waste of many different animals as a source of food, or also for providing brood balls for larvae, which either barrow in the ground or live in chambers.

[Source: https://scitechdaily.com/crappy-news-for-the-dung-beetle-and-agriculture-greenhouse-gasses/amp/*]*

These groups of insects are classified in the superorder Endopterygota, in the order Coleoptera. We can easily distinguish them from many of the other insects due to their front pair of wings which are much harder (wing-cases) and these are referred to as elytra. This order is one of the largest orders with approximately 400,000 species making up for almost 41% of all insects that have been described and 26% of all described animal life forms; with new species still being discovered.

The Curculionidae Is the largest family within this order having approximately 83,000 species within this order. They can interact with the ecosystems in several ways, seeing as they are found in the majority of the habitats available except the polar regions and the sea. These interactions include: eating other invertebrates, breaking down plant and animal debris, and often feeding on fungi and plant. Some species have proven to be dangerous agricultural pests, such as the potato beetle that is found in Colorado; however, there are others such as ladybugs that have been used as pest control insects, eating thrips, scale insects, aphids, and other sap-sucking insects that damage crops.

The majority of the Beatles found globally have been considered as important dangerous pests of stored products and agricultural plants. They end up attacking all the living plant parts, in addition to wood products, greens, and processed fibers. Wood- boring beetles and scavenger beaters have proven to be useful as recyclers and decomposers of organic nutrients. Predatory species, such as Lady beetles, have proven to also be biological control agents of scale insects and aphids.

10.5.8 Odonata

This order includes insects that can fly such damselflies and dragonflies. Just like other insects that can fly, for example, Hymenoptera, Lepidoptera, beetles, and flies, they were able to develop in the Mesozoic era. The prototypes, 325 mya, the giant dragonflies of the Carboniferous, are no longer classified as Odonata; however, they are classified under Meganisoptera or Protodonata.

Most damselflies and dragonflies are considered insects that are beneficial due to the fact that they feed on certain small flying insects such as mosquitoes. They are however regarded as pests by beekeepers because they can also catch and eat honeybees.

Figure 10.17: Odonata is an order that includes dragonflies and damselflies. Dragonflies feed on small flying insects such as mosquitoes, and their availability is very beneficial to outdoor enthusiasts and gardeners. In addition, their presence benefits the environment by making sure that humans reduce their pesticide use for killing these stubborn insects.

[Source: http://indobiosys.org/indonesian-biodiversity/taxon-images/odonata-dragonflies-and-damselflies*]*

In certain European countries, dragonflies have been classified as a threat in the poultry industry, due to the fact that they carry any lead to the transmission of a parasitic flatworm referred to as *Prosthogoniums pellucidus*. They feed on flatworm cysts, causing them to become infected with the parasite. After they ingest these cysts, these cysts turn into adults within the dragonflies where they end up spreading to birds that feed on these infected dragonflies (mainly poultry). These parasites affect the birds' respiratory system, killing them.

These are just but a few of the endangered insect species. The main reason I have included the importance is for you to understand that there are many reasons why they could be declining. We as humans consider a number of insects as pests and end up killing them with the use of insecticide, as well as other various techniques. Some other insects have led to the damage of crops due to feeding on their wood, fruits, leaves, or sap. Other insects are parasitic and may also be disease vectors. There are however others that are important for performing some complex ecological roles, for example, blowflies, which help by consuming carrion; however, they may also be carriers of diseases. Of the most essential are insect pollinators that are very important in the lifecycle of many various flowering plant species, which the majority of the living organisms, not excluding humans, are dependent on in one way or another. Without insect pollinators, there would be the devastation of the terrestrial portion of the biosphere. It is also true to say that many insect species are ecologically beneficial due to predatory behavior, and there are some with the same behavior and have proven to have a direct economic benefit. Honey bees produce honey and silkworms produced silk and there has been domestication of these insects by humans. 80% of the world's nations consume insects as food, by people that are in approximately 3100 ethnic groups. Insect biodiversity is highly affected by human activity.

10.6 REDUCING INSECT DECLINE

Changes in the climate, overexposure to pesticides and loss of natural habitat are just a few of the factors that have contributed to insect decline, including some species that were previously common of bees, Beatles, butterflies and flies among others. There has been a disappearance of more than two -thirds of all populations of caddisflies in the past decade.

Figure 10.18: There are many ways to attracts insects. One of them is by providing them with favorable habitats that consist of flowers or herbs. For example, butterflies love nectar -rich flowers, and having this can attract many butterflies your way.

[Source: https://www.discoverwildlife.com/how-to/wildlife-gardening/how-to-make-a-butterfly-friendly-garden/*]*

There are many birds and also larger animals that depend on insects as a source of vital food, in addition to being important for crops as well as wildflower pollination. If we continue in this direction with the current rate of decline, there are many negative consequences that we as humans, as well as the planet and all that lives on it, will face. There are some practical things that we can do to reduce this decline, especially in urban areas.

10.6.1 Growing Plants

Bees ,as well as other pollinating insects, need flower- rich habitats which have all disappeared. We need to understand that it is up to us to make sure that they are replaced, especially in cities and towns, which are difficult environments for the survival of these insects.

Because the majority of us live in the cities, we can help by placing or growing plants in even the smallest spaces, such as a balcony or a window box. For those who have large pieces of land, they should make it a norm to plant flowers and other plants that can attract pollinators and allow them to thrive. For those who have gardens, they should fill them up with flowers and herbs which will also attract pollinators. It is also possible to grow vertical gardens in tight spaces on the rooftops or external walls.

10.6.2 Creation of Insect Sanctuaries

Figure 10.19: Considering the fact that we have chosen to destroy many insects habitats, insects sanctuaries have become a modern innovation that many people are using to attract insects to the cities and towns. These sanctuaries can act as a home for insects that do not have natural habitats, especially if they are well maintained.

[Source: https://www.independent.com/2019/04/03/cities-as-sanctuaries-for-bees-insects-and-birds/*]*

Seeing as habitat loss is also one of the major issues promoting insect population decline, another way we can counter this problem is by building or buying environments that bugs can thrive in, such as beehives, butterfly houses, and even ladybird towers. In places where there is not much space, using these environments or structures can help in encouraging beneficial insects, boosting of biodiversity and also providing a nice way for a close up observation of nature, especially by young children, in addition to improving the aesthetics.

10.6.3 Eating Organic Foods

When we talk about organic food, we are talking about vegetables and fruits that do not have any artificial fertilizers, seeing as these artificial fertilizers have proven to be very dangerous to insect habitats. Organic plants are grown using fertilizers that are all- natural, hardly ever containing any traces of chemical pesticides that always end up seeping into non-organic foods. This also means being aware of what we are buying in the supermarket and discouraging the purchase of non-organic foods.

10.6.4 Turning Off the Lights Outside

There are many different insects, including moths, that have an attraction to artificial lights, despite the fact that they have no idea that these lights could be the reason why they die. Not only do these artificial lights attract insects and lead them away from their natural habitats, therefore decreasing biodiversity, but also end up burning them to death.

Of all insect species, half of them are nocturnal, and that means they are highly dependent on natural light in darkness from the night sky and the moon for the guidance of their movements, reproduction, and even finding food.when we switch on too many of our artificial lights, these attract insects from the natural habitats, and we unknowingly end up disrupting their breeding patterns, as well as decreasing their chances of survival, and this happens with many insect species; some that we know and some that we do not know about.

10.6.5 Going Green in the Garden

The chemicals that we find in pesticides and fertilizers are neurotoxic, and the majority of the time is deadly for insects. In addition to this, these chemicals may end up making their way into sources of water, such as rivers or ponds, and end up in the ocean eventually. This is normally a hard topic to discuss because, despite the negative effects, they have helped in preventing large crop losses and therefore, in accordance with the World Health Organization, pesticides and insecticide will continue to be used in the agricultural sector for a long time to come.

Figure 10.20: In addition to water pollution, artificial fertilizers can degrade the soil, making it less fertile, therefore making the plants weak and unsuitable for certain insect habitats. In addition, artificial fertilizers can attract certain diseases, killing off plants or making them inhabitable.

[Source: https://medium.com/orgpick/organic-vs-non-organic-food-bd501516c195*]*

This does not mean that we cannot make the small changes that need to be made. In the household, gardeners should consider using products that are all- natural, such as compost that they make from organic waste, recycled leftover food, and even grass clippings, as alternatives that are friendly to the insects. In addition to this, it is possible to also just leave the insects alone and wait for the appearance of natural predators, which will deal with the stubborn insects for them.

There are also other solutions to these problems, however, there is a lack of a large scale vision, major institutional funding, and political commitment. To deal with the loss of habitat, scientists are pushing for the preservation of biodiversity hotspots, for example, nature- friendly agriculture, regenerating damaged ecosystems, and even protecting primary rain forests.

In addition, scientists agree that so as to deal with climate change, there is a need to cut down the deep carbon emissions, together with having to establish large and secure corridors and areas that are well conserved, all that will encompass many different tropical and temperate ecosystems, which should sometimes be created to preserve endangered insect species.

Some of the solutions that have come up when it comes to the use of pesticides include banning a number of toxins, as well as pesticide seed coatings; making sure that scientists well educate farmers, instead of having the pesticide companies educate them; and most importantly, having to rethink the practices that are used in agribusinesses. Some of these elements are already included in the plan for biodiversity recovery in the Netherlands.

References

1. Marshall Cavendish. (2000). Exploring Life Science, Volume 6. Singapore: Marshall Cavendish, pp880.
2. A. Khaliq, M. Javed, M. Sohail, & Muhammad Sagheer. (2014). Environmental effects on insects and their population dynamics . Journal of Entomology and Zoology Studies , 2 (2): 1-7.
3. A.P. Rasnitsyn & Donald L. Quicke. (2007). History of Insects. Berlin: Springer Science & Business Media, 1-65.
4. Adkins-Regan, E. (2013). Hormones and Animal Social Behavior. New Jersey: Princeton University Press, 1-34.
5. Berenbaum, M. R. (2010). Bugs in the System: Insects and Their Impact on Human Affairs. New South Wales: ReadHowYouWant.com, 1-64.
6. Brumm, H. (2013). Animal Communication and Noise. Berlin: Springer Science & Business Media, 6-91.
7. Dudley, R. (2002). The Biomechanics of Insect Flight: Form, Function, Evolution. New Jersey: Princeton University Press, 3-75.
8. Engels, W. (2012). Social Insects: An Evolutionary Approach to Classs and Reproduction. Berlin: Springer Science & Business Media, 6-77.
9. Gilbert, L. I. (2011). Insect Endocrinology. New York: Academic Press, 1-93.
10. Gordon, D. M. (2010). Ant Encounters: Interaction Networks and Colony Behavior. New Jersey: Princeton University Press, 1-75.
11. Hedwig, B. (2013). Insect Hearing and Acoustic Communication. Berlin: Springer Science & Business Media, 5-101.
12. Heming, B. S. (2018). Insect Development and Evolution. New York: Cornell University Press, 1-65.

13. Henrik Steinmann & Lajos Zombori. (2012). Dictionary of Insect Morphology. Berlin: Walter de Gruyter, pp.405.
14. Herreid, C. F. (2012). Locomotion and Energetics in Arthropods. Springer Science & Business Media, 1-73: Berlin.
15. Jan Christian Habel, Michael J. Samways, & Thomas Schmitt. (2019). Mitigating the precipitous decline of terrestrial European insects: Requirements for a new strategy. Biodiversity and Conservation , 28:1343–1360.
16. Kaner, E. (2002). Animal Talk: How Animals Communicate through Sight, Sound and Smell. New York: Kids Can Press Ltd, 6-38.
17. Klowden, M. J. (2013). Physiological Systems in Insects. Massachusetts, United States: Academic Press, 1-89.
18. Leather, S. R. (2008). Insect Sampling in Forest Ecosystems. New Jersey: John Wiley & Sons, 1-77.
19. Locke, M. (2012). Insect Biology in The Future: VBW 80. Netherlands: Elsevier, 1-79.
20. McGregor, P. K. (2005). Animal Communication Networks. London: Cambridge University Press, 9-84.
21. Morris, B. (2006). Insects and Human Life. Berlin: Berg, 1-49.
22. Noest, R. M. (2017). INSECT LOCOMOTION: CONTROL, STABILITY, AND OPTIMIZATION. New York: Faculty of the Graduate School of .
23. P. S. Cranston & P. J. Gullan, . (2010). The Insects: An Outline of Entomology. New Jersey: John Wiley & Sons, 1-91.
24. Prakash, M. (2008). Insect Ecology. New Delhi: Discovery Publishing House, pp368.
25. Price, P. W. (2003). Macroevolutionary Theory on Macroecological Patterns. London: Cambridge University Press, pp291.
26. Schowalter, T. D. (2006). Insect Ecology: An Ecosystem Approach. Netherlands: Elsevier, 1-95.
27. SERGEY YA. REZNIK, NATALIA D. VOINOVICH & NINA P. VAGHINA. (2009). Effect of temperature on the reproduction and development of Trichogramma buesi (Hymenoptera: Trichogrammatidae). Eur. J. Entomol. , 106: 535–544.

28. Shaik Javed & Raju Agurla. (2016). The Cardinal Traits of Insect Morphology and Physiology. New Delhi: Educreation Publishing, 1-53.
29. Shaw, S. R. (2014). Planet of the Bugs: Evolution and the Rise of Insects. Chicago: University of Chicago Press, 1-70.
30. Shields, V. D. (2017). Insect Physiology and Ecology. Norderstedt, Germany: BoD – Books on Demand, 9-55.
31. Shields, V. D. (2017). Insect Physiology and Ecology. Norderstedt, Germany: BoD – Books on Demand, 25-55.
32. Snodgrass, R. E. (2018). Principles of Insect Morphology. New York: Cornell University Press, 1-83.
33. Swanson, M. L. (2013). Communication: from Pheromones to the Internet and Beyond. Canada: FriesenPress, PP312.
34. Walter, G. H. (2005). Insect Pest Management and Ecological Research. London: Cambridge University Press, 1-13.
35. Wigglesworth, V. B. (2012). The Principles of Insect Physiology. Berlin: Springer Science & Business Media, 1-61.
36. Wyatt, T. D. (2003). Pheromones and Animal Behaviour: Communication by Smell and Taste. London: Cambridge University Press, pp391.

INDEX

Q

R

S

T